FINANCE DEPARTMENT TRAINING

CONTINUITY PLANNING

Preventing, Surviving and Recovering from Disaster

Ronald D. Ginn

Elsevier Advanced Technology
Mayfield House, 256 Banbury Road,
Oxford OX2 7DH, England

I

Copyright © 1989
Elsevier Science Publishers Ltd.
Mayfield House, 256 Banbury Road, Oxford OX2 7DH, England

British Library Cataloguing in Publication Data
Ginn, Ron
Continuity planning.
1. Management. Planning
I. Title
658.4'012

ISBN 0-946395-43-8

For my grandchildren, Thomas and Kathryn,
May their world be a kinder and safer place.

Ron Ginn is European Managing Director of a well known continuity planning consultancy.

After graduating he worked for IBM for some fourteen years holding executive appointments in both Europe and North America. His subsequent career included senior positions with Rolls Royce and the International Division of ICL.

Mr Ginn is a member of the American Society of Industrial Security, a Freeman of the Company of Information Technologists, and a Freeman of the City of London. He is a frequent speaker at conferences in both Europe and North America on security, continuity and disaster recovery planning subjects.

PREFACE

So many people have helped directly and indirectly on the preparation of this book that it is impossible to credit them all. I must, however, mention four of them. Norm Harris, the doyen of our industry who has over the years become a firm friend; Randy March, another such, who patiently helped me during my learning phase; Gina van Hardeveld, who gave up so much of her time to decipher my handwritten scrawl and type the text and Rob ter Mors of Elsevier without whose persistence the project would still be being put off until "I have more time".

My thanks to them and to all of the other friends who have read and commented on the book as it developed.

CONTINUITY PLANNING

CONTENTS

CONTINUITY PLANNING

INTRODUCTION

Back in the early 1970s, three people were working together on the development of a very large computer software system. They grew to know one another very well and to respect each others strengths.

One of the three was a vice president of the bank for which the system was being developed and during one late night business session he said to the others, "Do you know, I am tired of receiving calls from the data centre at any hour day or night, no matter where I am or what I am doing over some crisis or another ..... I think that I need some kind of management plan, a plan to help determine whether a situation is a potential disaster and to control it".

The three people concerned were Norman L. Harris, Edward S. Devlin and Judith Robey and out of that late night session emerged the backbone of all disaster recovery concepts in use today ....., the first schematic with ready computer backup sites identified, the forms and formats containing essential recovery information, file recovery methods and on, and on and on. There and then was the birth of a new industry — the DISASTER RECOVERY, now the CONTINUITY PLANNING INDUSTRY.

Norm, Ed and Judy conducted the very first disaster recovery educational event and have since been involved in the development of more than half of the plans in existence in North America today.

Over the years the necessity of disaster recovery planning has become increasingly evident. People have come to realize that it is insufficient only to have made arrangements for an alternate work place should they lose their own location. The trick is to get everyone and everything necessary to continue work to that location probably at two o'clock in the morning and possibly in the middle of a blinding snowstorm with no public transport available. That takes detailed planning. It has also become evident that insurance cover is no protection because operating efficiency deteriorates so swiftly after a major disaster (see Chapter 1) that a company can often be bankrupt before the claim is met.

For these reasons, more and more organizations turned to Harris, Devlin and Robey for help and the methodology which they have developed has been used by large and small organizations in every industrial and commercial market sector and, also, by government departments. They are now the three chief officers of Harris Devlin Associates who are the leading suppliers of disaster recovery and security consultation and services in North America today. Continuity Planning Associates is the company formed to provide these services, now based on over 15 years of experience, in Europe and elsewhere in the world. Of course, those initial plans were plans entirely concerned with surviving and recovering from a disaster to a computer centre and for many years the plans continued to be limited in this way. As the experience of the three industry pioneers grew, however, and some disasters were successfully encountered, it became evident that such limitation in the scope of a plan was in itself dangerous. Obviously, information processing is at the heart of every organization and its serious disruption or destruction will seriously effect or also destroy every operation. A plan for data processing is, therefore, essential and is often the first to be developed.

Experience shows, however, that such a plan is not sufficient to protect a company from the threat of business failure, of bankruptcy, through a major disaster. Loss of a headquarters building, a distribution centre, or any of a host of other vital locations can be just as destructive. Experience shows, also, that taking preventive measures to lessen the risk of a disaster occurring is usually cheaper and always far more preferable to experiencing one. And so the methodology and scope of the planning provided by Harris Devlin has been widened and expanded over the years to include disaster prevention and to be applicable to all departments and functions within a company or organization.

Unfortunately, there are many, more recent, arrivals in the industry who still supply plans based on the original Harris, Devlin and Robey concept and methodology. It is for this reason that the term continuity planning has been developed. In this new context a disaster recovery plan is one developed to enable survival and recovery from a disaster to a computer centre. A continuity plan in contrast, has two distinct phases, the first is activity designed to lessen the risk of disaster occurring, the second is that of developing a survival plan designed to ensure survival and recovery from a disaster to the total company or any part of it. In effect, the part of a survival plan which is applied to the data processing function is the old disaster recovery plan!

This book is about continuity planning.

CONTINUITY PLANNING

CHAPTER ONE

Why a Plan is Necessary

When Disaster Strikes

What would happen if your information processing systems, your headquarters building or some other major location or function in your company disappeared right now? If your answer is "nothing" then you should not have wasted your money in buying this book!

There are a few fortunate companies in such a position. I know of one such, a specialist commodity brokers. There are five dealers involved who each know their limited client list intimately. They work from about 10 a.m. to 3 p.m. each day from Tuesday to Thursday and the number of deals involved is small enough for the details to be carried in their heads. They have a small computer but it is only there as a show piece and for playing games' and the considerable profits which they make have been invested in office property and so there is always a spare office suite somewhere to which they can move should their normal building be destroyed.

Most of us are not so fortunate, however, and we are living in an environment of ever increasing risk of experiencing a disaster (especially in the areas of environmental or man-made, malicious disasters) in which the time available to respond and recover is being dramatically reduced.

One factor which contributes to both sides of this equation is the rapid growth of distributed data processing involving networks of both computers and terminals. One international company, for example, has a worldwide network of over 100 000 terminals processing on-line applications including just-in-time manufacturing. A large proportion of the information in the system at any one time has been produced within the system itself and never seen by any of the personnel. The company estimates that if the network is out of operation for more than twenty minutes it will have a devastating and costly effect.

Another factor is the increasing use of powerful personal, micro and mini systems. In many cases these are used by non data processing staff to process confidential or sensitive applications which require a high level of security. Such staff often have little, if any, knowledge of even the most fundamental information security procedures and the results of this would sometimes be funny if the danger posed was not so serious.

One company recently circulated a memorandum to all departments giving advice on elementary precautions to be taken should they be using a personal computer. On carrying out a check to see how effective the memorandum had been, they found one microcomputer user, following the advice to place his data diskettes in a secure place when not in the office, punched holes in his diskettes, rendering them totally useless, and then carefully filed them in a two ring binder. Another user, following the guidelines to take back-up copies of her data files, did — by photocopying the diskettes! She then dutifully filed the resultant, totally useless, sheets of paper securely under lock and key.

It is also worth remembering that many of these small machines, used by such untrained staff, are also used as terminals allowing access to the company network.

The effect of a disaster can be devastating. Industry-wide statistics show that essential company functions can continue for only four to five days after a disaster to its data processing operations. Even the most optimistic statisticians state that at between four and one half and five and one half days the company is reduced to approximately 50% efficiency and by the eleventh day it is down to 9%.

Most firms limit their planning to the old-fashioned area of DP disaster recovery. A recent survey of UK firms showed that 52% of the companies surveyed said their business would be damaged if computers were out of action for one day or longer. A sizeable proportion had no plans whatsoever for such disasters. That is the scope of the problem viewed simply in DP terms. When the problem is properly researched as the business problem of continuity planning an even higher proportion is likely to show itself to be completely unprepared.

Assessing the Risk

The risk involved is highlighted by a recent report of a comprehensive study which showed that of all of the organizations surveyed which have suffered a disaster which was only to their data centre and which did not have a survival plan, less than seven percent were still in business after five years. Most of them, in fact, survived for less than one-and-a-half years.

The conclusions of another survey, conducted by the College of Business Administration of the University of Texas are that most organizations are very dependent on computer systems to support vital business functions. The survey indicates that 85% of the organizations are heavily or totally dependent upon computer systems and that financial and functional loss increase rapidly after the onset of failure. The survey results show that, within two weeks of the loss of computer support, 75% of organizations would have reached critical or total loss of functioning, loss of revenues and additional costs rise rapidly and become substantial as the loss continues. One financial industry respondent stated, "We would be out of business after one week." A manufacturing industry respondent reported, "After 25 days of loss, our chances of coming back as a corporation are about 20%." The overall survey results indicate that these are typical and not isolated cases.

In North America, government realization of the risk involved has led to introduction of Federal legislation which mandates that all financial institutes must have at least a survival plan which should be externally audited and reported on.

Tales of Woe

However, as already discussed, disasters to computer systems are not the only ones which can force a company into liquidation. One example, well known in the industry, is of a spectacular fire which occurred one Thanksgiving evening at the headquarters of a major US bank. The building was completely destroyed by a fire so intense that it melted safes which were rated as being proof against fire for four hours.

The bank had a survival plan covering all of the business functions which had been identified as crucial to its continuance and, in consequence, was back doing essential business after three days — three days which were, incidentally, public holidays. The plan enabled them to relocate essential staff into pre-determined alternative

Fire can wreak havoc

locations and perform the hundreds of detailed tasks necessary even in a vacation period.

Analysis of this incident reveals some important facts:

— This was an example where no significant data processing equipment or computer hardware was involved. It was just prior to the spread of word processing and microcomputing.

— Even so, the bank senior management have stated that if they had not been able to restart so quickly by using the survival plan then the bank would have been out of business, bankrupt, within four or five weeks.

— The bank was able to rebuild the headquarters building even better than before because the insurance cover, backed by the detail in the survival plan was extremely comprehensive. The problem was that, without the quick restart, this money would not have been approved and paid fast enough to save the bank from bankruptcy.

— The bank had not developed a full continuity plan in that they had not performed a Security Review. If they had, the probability is that the fire would never have started.

So the effect of a disaster can be devastating, but what is the danger? What are the disasters which threaten us? They fall into three main categories which are accidental; natural; and man-made and malicious.

Accidental disasters include such events as fire which, statistically speaking, was until recently the biggest single risk facing most organizations. It also includes disasters which occur through such things as errors and omissions or, for example, maintenance accidents. There was a case, recently, where on completion of some electrical work the supply was incorrectly reconnected which resulted in the melt-down of all of the computer mainframes in a financial institution.

Natural disasters include flooding, blizzards, electrical and wind storms and earthquakes. Any of these, if severe enough, can destroy a facility or prevent access to it for long enough to create disaster. In this context, it is worth remembering that disaster is not necessarily caused by the destruction, the total loss, of a facility or location. Not being able to enter it can cause total disaster if it lasts for long enough. A flood, evacuation of the area because of poison gas thrown off from a neighbouring factory fire or because of a terrorist bomb threat to a nearby building, or even industrial action, can deny your entry to your building.

A striking example of the effect of an 'outside' disaster is provided by a recent fire in a public telephone hub rerouting station which destroyed over 350 000 voice and data lines. The effect was dramatic. With phone lines dead, commerce virtually stopped. Airlines grounded flights, automatic teller machines couldn't execute transactions, banks couldn't verify credit ratings, and high-priced sales reps found themselves pulling duty as messengers. Branch offices that rely on telecommunications to send data to headquarters rushed employees with cartons of paper to central offices to ensure data were entered into central computers. Cellular phones graduated from status symbol to necessity; anyone who could get their hands on one did so.

The failure of the telephone sub-station wrought crushing consequences for the area. Although corporations have yet to assess the full impact of the fire on their businesses, it is evident that its effects were devastating. The telephone company itself says it lost over £500 000 a day in business from local calls. The general manager of the local agency of a national insurance company, estimates that the fire cost the agency approximately £250 000 in lost business and severely constrained its ability to service policyholders. A travel agency has filed a class action suit against the telephone company for an undisclosed amount in damages. Without phone service, the agency asserts, it was effectively "out of business". Ten other area companies have cast in with the travel agency and the agency's lawyer says that two of those companies reported substantial losses: One reported losing £50 000; another £20 000 a day for 14 days.

Even corporations with private satellite networks, dialup modems, and microwave dishes as part of their survival plans suffered greatly and, at the time of going to press, it is still to early to say how many companies may disappear as the result of this incident.

Man-made disasters cover a wide variety of malicious activities including withdrawal of labour, industrial espionage, sabotage, international terrorism, computer 'hacking' and computer related fraud. The danger involved in some of these is reasonably obvious and companies are often able to judge their degree of exposure without too much difficulty. Others are fairly new, in many cases rapidly growing, and so the exposure may not be so readily appreciated. Among these are industrial espionage, hacking, terrorism and fraud.

Industrial Espionage

Many people treat the idea of industrial espionage with hilarity. They think that it is a huge joke that only occurs in sensational fiction. A glance around the electronics shops in any major international airport should dispel that illusion. These shops do not stock the large range of sophisticated bugging devices on display unless they sell them and people buy them to use them. The British Post Office was recently awarded millions of pounds in damages against an American company which had stolen from them the secrets of a new device which they then marketed.

Hacking

Computer 'hacking', which can be defined as the unauthorized entry into a system, was, until a few years ago, mostly done by young people whose motivation was to beat the system. Usually using a personal computer and a modem to connect it to the telephone system they would be satisfied to break into a mainframe and leave some form of message to prove that they had done so. Unfortunately, some of them have developed into criminals or even political activists. They are formed into clubs which pass information from one to another and are becoming increasingly sophisticated.

Dial-back systems, encryption and other methods have all been used with varying degrees of success to beat the hackers but soon after most of these are developed the hackers find a way around them. Two of their techniques, which come under the general heading of 'trawling', are becoming more widespread. These are:

* interception and translation of the microwave

transmissions from terminal screens. With modern but relatively inexpensive equipment this can be done up to a considerable distance from an installation.

* bugging an ofice or computer installation to pick up the noises emitted by a typewriter or many types of printer. These noises are unique for each machine and can be translated by a computer program to duplicate what was being typed or printed.

A company in Germany recently received a phone call in which the caller stated that he had hacked into their system and planted a 'computer virus' which, if activated, would cause considerable damage. He offered to tell them where it was in return for a reasonable sum of money. Eventually they paid him and he revealed the hiding place of a few lines of programming which would have caused the gradual distortion of all of their inventory records.

They calculate that this would have bankrupted them in about nine months.

Some hacking does have its humorous side. There is the case of the hacker who broke into a Head of State's private file and drastically but subtly changed a major speech just minutes before it was printed out for him to present. Unfortunately neither the perpetrator nor the results are known.

It is important to realize, however, that a substantial proportion of hacking occurs from within a company. Sometimes this is done by employees accessing files, which they are not authorized to see, out of curiosity. More and more frequently it is done for reasons of sabotage or to commit fraud.

Terrorism

International terrorism is another area of growing danger because the terrorists fully realize that the increasing inter-reliance of our society has changed the nature of the threat which they can pose. Much of the functioning of our heavily industrialized societies is becoming more and more concentrated on a decreasing number of critical locations or processes both in countries as a whole and in individual organizations.

The exposure from terrorism is often not fully appreciated because the total costs resulting from an incident may not be known for some time. They will, therefore, often bear little relationship to the figures published at the time when the incident occurred. For example, one bombing of the International Division headquarters of a company which resulted in no injuries had an initial estimate of cost, based on property damage, of some £50,000. Once it was possible to look at the add-on effect of staff losses through fear (44 people), the cost of recruiting and training replacements, lost production time etc. the true cost was calculated to be £500,000 — ten times as much.

In the view of speakers at an Anti-Terrorism, Espionage and Crime World Conference and Exhibition held in Washington D.C.:

* Computer centres are increasingly becoming a target of terrorist groups who equate computers with imperialist control. In the two years 1984/85, the latest that figures are available at the time of writing this report, some 50 attacks against computer centres by terrorist groups were recorded.

* Commercial companies will be attacked more frequently since terrorists are realizing that such attacks create more terror and can have a greater impact on the economy than military or government ones.

Growth in Fraud

Another major area of relatively new and rapidly growing exposure is that of computer related fraud. Until recently the greatest exposure to most organizations, statistically speaking, was fire. Some experts believe that this has changed over the last few years to computer related fraud.

The extent of this problem, quite frankly, is not known but there have been many sophisticated guesses. One puts the amount involved in one year, 1984, in the UK alone at over four billion pounds. Another is that US bank losses to their own employees in such frauds amounted to ten times their losses to conventional bank robbery in 1986.

One factor preventing an accurate assessment of the

problem is the low rate of convictions from such crime. International criminologists believe that only one person in thousands committing computer related fraud is ever convicted. They say that this is because only some 1% of such crimes are detected, of that only about 12% is reported because of reluctance by organizations to admit to having been victims and only 3% of that are convicted because of the wording and conflicting nature in many countries of the laws about computer fraud.

A second factor is the nature of the people who commit these crimes. They are usually senior or middle management or long-service employees. People considered to be above suspicion. These are the people with motive. Their careers and, therefore, their incomes have reached a plateau but their expenses; the mortgage, education fees, etc., continue to rise. They also have the knowledge and the opportunity both of which are, usually denied to more junior employees.

A third factor is the combination of the increasing complexity of computer systems, the time pressure to complete the development of systems and lack of knowledge by systems developers, operators and users of the risks and the precautions which should be taken.

Based on the number of occurrences of demonstrated and published frauds, the following classification can be made:

- the most frequent type of fraud is the one committed through manipulating input data; followed by:
- incorrect additions to master data;
- manipulation of data in correction accounts
- manipulation or destruction of output
- unauthorized modifications in job control language
- unauthorized additions to programs
- unauthorized additions to or mutations in operating systems.

Only very few cases are known of advanced computer fraud, most are based on manipulation of input data. However, advanced frauds are the most sensational ones, for they contain a romantic element and therefore get a great deal of attention.

11

The following case is a spectacular example of an advanced fraud based on the manipulation of rush orders.

The company concerned was in the business of wholesaling cosmetic products and was a subsidiary of an American company. They had gone to great lengths to install internal controls in the order processing system, for example:

- Development, programming and program maintenance was in the hands of the D.P. staff of the mother company. In this way the applications could not be influenced internally.

- The operating system of the computer being used was kept under such surveillance that manipulation of it was practically impossible.

- The system used in the fraud was the on-line order entry and invoicing system which had been created entirely in accordance with all standards imposed. It included any required programmed checks, such as automatically and continuously numbered forms for delivery and invoicing, plausibility checks, access codes, etc.

- Delivery planning was conducted on a seperate mini computer. There were no built in checks for completeness of the invoices. The reason was that such checks made no sense because a number of customers, by appointment, came to collect the goods themselves.

- The forms received back after a delivery were re-entered into the computer in order to process the right prices (depending on quantities, customer type, etc.), to generate the invoice, to update the stocks and to create statistics.

- The auditors kept a close watch on the cancelled delivery tickets because they felt, with good reason, that abuse of them, in cooperation with a buyer, offered a good opportunity for fraud.

- All required checks were built into files, access procedures, and even into specific records in the files.

On the basis of these facts one may assume, as the company concerned did, that everything had been done to make the system crime-proof. This, unfortunately, was not the case.

The sales manager of the company was ambitious, hard working and did his job satisfactorily. He knew very little about computers and therefore was not capable of directly influencing processing. What he did know, however, was that the computer could not process rush orders. He knew this because he had the responsibility, if a customer called in the afternoon, to write a manual delivery form and to have the goods loaded and delivered before closing time. This circumstance offered him the opportunity to commit a fraud amounting to nearly one million pounds.

The manually created delivery forms were the same as the normal computer forms, but as they were not printed, they remained unnumbered and therefore unchecked. The forms returned and signed by the customers were entered into the computer for normal processing. The sales manager's trick was to call a number of his small volume customers and offer them a special discount. His story was that the company's president had decided to make this special offer to trusted customers in order to raise sufficient money for advertising campaigns which the mother company, not understanding the local circumstances, would not allow. If sales were made to this customer, payment would have to be made to the confidential lawyer of the company, so as not to make the mother company suspicious. The customer was requested urgently not to discuss this with the normal sales representative, because this was a confidential matter which should be given the least possible attention. The president had entrusted everything to him (the sales manager) and did not want to be charged with the execution himself.

Some eight customers accepted the story and placed orders. The sales manager waited until the afternoon and then issued rush orders which were sent to Despatch for immediate delivery. The carriers were ordered to send the delivery forms received from the customers to himself, for "tax reasons".

The customers paid, without official invoice, to a lawyer who had been told the following story. The sales manager was charged with dumping excess stocks. For tax reasons these sales had to be paid for into a neutral account with an impartial agency, whereupon the money was withdrawn by the sales manager.

The lawyer concerned bought this story and so the fraud could be committed with all forms handed to the sales manager so that no direct evidence could circulate within the company. This situation went on for six months, during which the sales manager regularly collected his money.

Of course, stock discrepancies occurred which drew the attention of the management and eventually a high level meeting was convened to which the sales manager was also invited. The general feeling was that the drivers were stealing from stock and the security department of the company was ordered to keep a close watch on the drivers. The sales manager felt even more secure because of this conclusion and increased his fraudulent activities. He even atrracted another two customers.

On an unlucky day for the sales manager, one of his customers called with a problem. The sales manager was absent and the customer let himself be put through to the presidents secretary who made a note of his inquiry about the inexpensive goods and passed it on to her boss. After some investigation, the latter discovered what was going on.

The following are some other frauds of interest:

— A purchasing manager created three suppliers with self-invented names and the addresses of girl friends as office addresses. He had forms printed and created invoices on his company for parts, advertising material and printed matter. He forwarded the invoices to his company and then OK'd them, certifying that the goods had been received. The false invoices were entered into the company's computers and the payments were made in time. These payments were made into accounts which the purchasing manager had opened especially for that purpose.

— An office clerk in a cash and carry radio and TV shop filled in codes for free replacement components on delivery tickets although, normally, these components had to be paid for. In this way the office clerk withdrew, in cooperation with a customer, some £25 000 worth of components from his employer.

— The manager of a large chemical company made use of the possibility that he could change master files. He attributed to a number of customers a high discount code which was meant exclusively for inter-company deliveries. Delivery forms and invoices were created by means of the computer system and the buyers involved enjoyed the very high discount. The manager received his share from the customers in cash.

So, these are some of the things that can happen, that do happen, and the traumatic effect that it can have on a business. Building a market is a matter of years of effort while losing a market can be a matter of minutes! Luckily the picture is not one of complete gloom and doom, we have Continuity Planning.

CONTINUITY PLANNING

CHAPTER TWO

Minimizing the Risk — The
Security Programme

Learning to Cope

Why isn't this chapter called 'Prevention'? Quite simply because prevention is not a practical possibility particularly when considering natural or environmental disasters. I suppose that if you could build a 'Fort Knox' type complex in the middle of a mountain you might get close to it but I wonder how many people would work for you.

What is necessary, practically, is to reduce the possibility of experiencing a disaster to the lowest possible level with the money available to take preventive measures. It also means being in a constant state of awareness of changes to circumstances and conducting a regular, thorough review of all preventive measures.

In the area of man-made malicious events, the object is to take steps to ensure that you are not an easy target. Someone intent on doing damage or stealing is not going to choose a location which is obviously difficult but, rather, one which appears to be the most slack in its protection. Most internal fraud or theft starts as a seized opportunity which is then systematically exploited. Remove the sudden opportunity!

All of this means, then, a comprehensive security programme followed by an ongoing system of security reviews and audits. In this chapter we will review the basic requirements for the security Programme and in the next security audit and review.

One of the first requirements is a proper understanding of what is meant by Security, the areas which should be included and the methods which should be employed. There is often a difference of opinion about this even within an individual company.

In part, these differences seem to arise because of both the rapid expansion of the areas of an organization which are exposed to risk because of technological advances, and also the concentration of sensitive, intangible assets, such as information, into relatively easily accessible areas. It also arises because traditionally, the risks faced by most organizations have come from without. There is now a growing danger, often unrecognized, from within, which needs to be protected against.

Security is all about the protection of assets, whether these be tangible such as money, buildings and inventory or intangible such as data, information, market share or employee safety. Some people consider such protection to be adequate if access to the assets is restricted and, of course, such restriction is important. Access must be allowed only to those who really have an organization or business need for it.

One interesting aspect of the need to restrict access to such items as computerized data is that a very large percentage of computer data loss is attributed to errors and omissions. Estimates range from 50% to 85% of the total losses. By restricting access to the various realms of data to those who have a justifiable need, the risk of someone accidentally entering an unfamiliar realm, and doing damage because of that unfamiliarity, can be minimized.

Restricting access to the assets is, however, only one half of the story. The other half, which is so often ignored, is that there must be accountability for all authorized access. If staff must account for their use of the assets then this becomes a strong deterrent to any wrongdoing.

Introducing Security

Suppose I sit two hundred people down in a hall and give five of them a key any of which will open a box on a table in front of them. I then tell them that there is a fortune in the box and that I am going to switch all of the lights out for an hour. I will guarantee that the money will be gone at the end of the hour but who has taken it? It is probably one of the five — I restricted access — but which one? There is also the possibility that one of them lost his key or had his pocket picked.

Now I will change the scenario slightly. Now in the box is an electronic device which will tell me not only which key opens it but at what time. This time when the lights come on the money will still be there because not only will the five not take it but they will make certain that their keys are kept safely.

Facilities should be in place to log all authorized accesses to assets; these logs must be reviewed on a regular basis and personnel informed that this is taking place. If accountability logs are produced but nobody reviews them and personnel realize this, then the deterrent is considerably weakened.

There must, then, be a company security policy which covers both restriction and accountability and another requirement of a good security program is that this must be coordinated.

The two basic overall areas of security are physical security and data or information security. These two areas are highly interrelated and dependent on each other. Yet in many organizations, even where both are addressed, they are the responsibilities of different people often reporting into completely different areas of the organization and within different management structures. A little consideration, however, shows that both have many areas in common.

Management objectives, policy, training and implementation are all aspects in which the responsibility for administration of the two areas should be combined or at least highly co-ordinated. If they are not, it is quite possible that a deficiency in one area could easily render the other ineffective.

During the development of the policy there must be good consultation with staff and a sensitivity to their feelings and requirements. The adoption of security measures, such as the two basic concepts of restriction and accountability, can sometimes lead to discontent within an organization when restriction is introduced where little or no restriction has been enforced previously. Long term employees can become very disturbed when suddenly they are told they can no longer access assets because they have no 'need'. All manner of protests about loss of confidence and trust may arise. For this reason and many others, top management support is absolutely necessary for any security programme.

Any security system will only be effective if the people to whom the security is applied are cooperative and contribute to the effort voluntarily. This can be achieved in many ways, but one of the ways which can help is to introduce the security measures as measures which protect the people involved.

Security should be seen as a benefit to the person rather than a hindrance. For instance, a card entry access system can be presented as having one aspect of providing more personal security or protection to the people within the organization. Maintaining the confidentiality of passwords or the propriety of access cards can help prevent the person from being blamed for another's deeds.

There will always be cynics, but the more people there are in the organization who are working for security rather than working to circumvent it, the more effective the security system will be.

The policy must also provide for a proper balance between the technical facilities and hardware measures and the accompanying administrative procedures.

The finest technical facilities and hardware will not be effective if not installed, implemented and maintained with adequate administrative procedures. A card entry access system will not be effective if, for instance, proper procedures have not been established to control the issue, granting of authority and retrieval of the cards associated with the system. Formal training in the use of the system and familiarization with the access authorization granted must be established. Logs produced by the system must be reviewed regularly and evident misuse of the system challenged. These are the types of administrative procedures which must be built around every technical facility. Where no technical facilities are available for protection, of course, the administrative procedures alone will have to ensure the protection.

It should not be forgotten that staff, too, are a vital asset and the security program should include measures for their protection. There should be adequate emergency alarms in place, proper emergency procedures developed and — a point so often neglected — these procedures must be practised on a regular basis.

The programme must also provide adequate measures for data or information security. It is obviously useless to have comprehensive protection of access to the computer system if you do not also protect access to the information the system produces whether in the form of printed matter, microform (microfilm or microfiche) or magnetic media.

To accomplish this in an effective manner, both from the standpoint of money cost and man-hour cost, the data or information should be classified for various levels of sensitivity, criticality and confidentiality.

Guidelines and procedures must be developed defining how data in each classification should be handled. This process should help to reduce the effort required to apply security measures to the data. Of course, the total amount of data or information involved must also be considered, there will be a level below which the effort to classify exceeds the effort to secure all data in like manner. Reason should prevail.

It is important to remember that assigning the classification of the data can only be done by the 'owners' of the data. The owners would be the department managers whose departments either create the data or are the primary collectors or users of the data. They are the ones in the organization who have the best knowledge of the sensitivity, criticality or confidentiality of the data.

Very often when organizations automate their information processing, department managers who had overseen the security aspect of the data or information previously, suddenly abdicate that responsibility to the data processing department. Data processing should be responsible for providing and maintaining the security required for the data in their role as 'custodian'. However data processing cannot possibly be familiar with all of the various aspects of required security on all of the data they process. It remains the proper responsibility of the owners to determine the security needs of the data.

The Need to Communicate

Once the security programme has been developed, the safety and security measures identified and the policy finalized, the next consideration is that of adequate communication.

Communication of the safety measures which have been adopted is often necessary to people outside of the organization who may, for example, be involved in an emergency.

The reason for this is best illustrated by the case of a company which installed breakproof window glass to prevent illegal entry. They did not inform the fire brigade and, when a fire occurred, a fireman tried to gain entry by breaking a window with a sledge hammer which rebounded and did him a severe injury.

Adequate thorough communication within the organization is always necessary and this needs to be not only about policy and methods but also about the reasons why. Unless the people who are responsible for implementing security have a thorough understanding then even so-called professionals are likely to fail to do the job properly.

One obvious example is that of a visitor waiting for an escort after the security guard or receptionist has carried out log-in and identity card issuing procedures. Often, if the visitor asks to visit the washroom while awaiting the escort, the guard will allow access through the security barrier to a washroom some corridors away. The visitor then has unescorted access to the building and often can even take along a briefcase.

These, then, are the basic requirements for a security programme but the overall consideration in developing it must, of course, be economy. As businessmen we must demand that our security programme be the same as the other operations in our organization, cost effective.

In establishing the program, the exposure of the organization to various threats should be identified. The exposures should be graded according to the potential damage to the organization and to the probability of occurrence of the threat. Then a reasoned decision can be made to adopt or not to adopt (and thereby accept the risk) security measures to cover the exposures. It can also be established to what level expenditure should be made to implement the security measures adopted in a cost effective manner. One thousand pounds should not be spent to protect assets worth one hundred pounds, and threats with a probability of 1 in 50 should be addressed before those with a probability of 1 in 1000.

CONTINUITY PLANNING

CHAPTER THREE

The Security Audit and Review

Being Systematic

Once the security programme is in place it should be subject to a regular system of audits and reviews. One obvious reason for this is to make sure that it has been updated to accomodate changes which have occurred to the organization, locality or business. Another is to make sure that the policies are properly understood and are being followed.

The receptionist knows, for example, that when the maintenance man from the copying machine company comes she has to verify his identity and check him against the list of authorized men from his company. She isn't going to go through that stupid formality for Charlie, though, is she, he has been coming to maintain the machines for three years and, anyway, he has taken her out once or twice. She doesn't know that he has been fired, while she was on vacation, she hasn't looked at the list of authorized personnel and so doesn't know that he has been removed .....

Whenever possible the security audit should be conducted by experienced external people. After all, as with all audits, you want it to be objective, unbiased and conducted without fear of repercussions from senior staff who may not agree with the findings. You will also want it to identify possible areas of wrong doing. The security programme being reviewed will have been largely developed by your own staff. In one such programme one of the staff concerned managed to include a nice little procedure in the networking security measures which enabled him to considerably inflate his overseas bank balance before he was discovered by the external reviewers.

Another reason for using external reviewers is that the internal staff who have developed the security programme will be hampered because they are part of the company and so think in the conventional company way. For example, I have a friend who is a senior manager in an organization which is extremely proud of its security procedures. I once teased him by telling him that I knew of a loophole in their access security and, when he wouldn't believe me, demonstrated it by being in his office waiting for him to arrive the following morning without having been officially admitted to the building. This has developed into a game between us in that every once in a while he will telephone to say that they know how I do it and have closed the loophole, at which I show them once again that they have not.

I am now going to spoil the game by telling how I do it. Their access security is based on the premise that all visitors will arrive either by public transport or by car. Cars are parked in the company's garage under the building and the drivers can only leave by one entrance which puts them into the entry security area along with arrivals by public transport. They also work on the premise that all of their employees will obey the rules and so, when those that travel by bicycle arrive, they will place these on the racks in the basement adjacent to the carpark and then come back out to go through the entry security area. I ride in on a bicycle, place it in the rack and then walk around to the back of the basement to the building elevator. Their problem is that they cannot conceive of a visitor coming on a bicycle!

The basic method for conducting a security audit is by holding in-depth interviews with personnel to establish their knowledge and acceptance of the security policies and methods and how well they are implementing them.

This again, illustrates the advisability for the audit to be performed by experienced external people. It is vital that the people being interviewed will have the confidence to confide their feelings and fears and it is unlikely that the necessary atmosphere of trust can be generated by people also employed by the organization particularly when they are in senior or management positions. It must also be appreciated that conducting such an audit to identify weaknessess in overall security requires a special kind of expertise. The people qualified to conduct such audits will not always be experts in the details of how to solve the problems which they identify but they must know where to advise their client to go in order to obtain the solutions and then be able to help to evaluate the proposals. This is important because a practical audit report must be realistic and contain solutions or recommendations for each of the points raised if it is to be useful and not just a theoretical consultancy exercise.

How to be Effective

To ensure that the review is effective several things must be done. First, there must be thorough consultation with the employees and their representatives, works councils, unions and so on. As already stated, the review cannot be effective unless the staff are prepared to cooperate fully and this will not happen if they are not properly involved from the beginning.

Second, interview questionnaires should be developed, documenting all of the points that must be investigated and designed so that the questions generate not simple 'yes' or 'no' answers but a discussion. It must be realized, however, that, in this type of audit, such questionnaires are not exhaustive but are merely guidelines, 'aides memoires', to ensure adequate coverage of the subject. The interviewer must use his or her experience and powers of observation to generate the ad hoc, detailed and probing extra questions which each individual situation will require.

Third, the interviews should be conducted not only with the people who are providing the security but also with those for whom the security is provided. The interviewees for whom the security is provided should always be asked if they have any concerns about security and if they know of areas where security might be improved. Sometimes security providers can overlook organizational security needs.

Next, as many of the points as possible should be discussed with more than one individual in separate interviews since a single individual may have an inaccurate understanding of the situation, or be trying to portray a better than actual circumstance. This of course, requires expert cross referencing during the report generation.

I am now going to describe in general terms each of the major areas which should be covered in a security review. Appendix 2 contains an in-depth list of topics for each area for those readers who need more detailed information.

Building and Site Security

The first major area to be reviewed is that of building and site security. The review should consider the implications of such aspects as geographic location and site surroundings; the contribution of layout, floor plans, type of construction and building materials to security and safety; the reliability of utilities; the ability and procedures of the security staff; the presence of terrorist threats.

It is generally not economically feasible to eliminate the possibility of terrorist attack by someone willing to sacrifice himself to accomplish the attack. The only thing that can effectively be done is to make such attacks difficult to accomplish thereby slowing an attacker down. This will hopefully minimize the possible damage and allow time to bring in appropriate assistance.

The security and safety of any room housing a computer system or office automation equipment should be reviewed using much the same points as those used in the building review. However, there are special considerations for the computer room which should be kept in mind, such as the higher level of access security required; clean environment needs; the need to protect valuable equipment and personnel during fire suppression; electrical power protection; protection from possible water damage and control and supervision of contracted service personnel.

Watching People

Review of the 'Human Resources' — or Personnel Security measures is also important since people can be one of the weakest links in a security programme. Sensitive and critical areas such as data processing must have reliable people. Thus the initial screening and employment practices, as well as training and termination procedures used by the personnel department are important and must conform to changes in the privacy laws and other legal restrictions.

Care should be taken to address the human resource side of fraud prevention by such methods as ensuring key personnel take sufficient vacation at one time so that their job must be performed by someone else; and by random checking for sudden changes in lifestyle. The vetting procedures for temporary personnel and suppliers' personnel should also be examined. It is amazing how many companies have good procedures for hiring permanent staff and yet allow temporary bureau staff to work without being vetted even in sensitive positions such as secretary to a director.

Access Control

The physical access control of the data processing network control centre must be reviewed as must the formal administration of access to data. Aspects including the access authorization process, maintenance of current controls, review of needed improvements, and follow-up of possible security violations or misuse of the data processing system are vital.

The control of the use of and the security procedures related to micro- or personal computers and word processing systems is another vital area. It is important that the access of these subsystems to central data processing be closely controlled. The technical difficulty of controlling data on these systems dictates the need for comprehensive policy and administrative procedures.

Being Consistent

As already discussed, in order to have an integrated, comprehensive security programme, there has to be central direction in the form of policy and guidelines. These could be reviewed for adequacy and currency. In spite of the existence of central policy, that policy may not be followed at the operational level of the organization. There must be a consistency between central policy and the procedures used at the operational level.

The adequacy of EDP audit in the organization must be looked at. Auditors should be involved in the development of security systems, application programme development, change control and most other day to day data processing functions. To do this the auditors must have the training and ability to work with computers and computer systems.

Finally the adequacy of all the technically-based controls for data security should be reviewed. The scope should be quite broad, covering the aspects of quality assurance, data base management, operating systems programming, application systems programming, and network/communication systems programming. Controls should be reviewed for the areas of systems development, operating procedures and change control applicable to all of these aspects. Dual control and accountability are very important.

CHAPTER FOUR

The DP Insurance Review

Insurance is a Minimal Measure

A common misunderstanding by people who have no experience of disaster is that all that is necessary to survive is insurance. This can be a dangerous illusion.

The operating efficiency of an organization deteriorates so rapidly after a disaster that in many cases a company is bankrupted before the insurance claims are met. The only people who benefit are the creditors and shareholders.

This does not mean, of course, that insurance cover is not necessary, the best of survival plans is of no use if the company cannot afford to implement it.

One of the problems which we often come across is that of inadequate data processing insurance. Many data processing managers do not understand insurance and have no experience of the effects of disaster and it also true that many insurance departments have little knowledge of the special requirements for data processing. In consequence many insurance policies for data processing contain serious omissions or exclusion clauses which can often be successfully negotiated at no cost.

When embarking on such negotiations it is important to remember that insurance companies exist to provide a service and must be made to do so. Unfortunately, some of them, sometimes, seem to need reminding of the fact that their customers needs must be paramount.

Points which should be considered in a review of data processing insurance are:

1. Equipment

Care should be taken that equipment is covered at replacement cost valuation and that the limits of liability are at an adequate level. Since the replacement cost of computing equipment generally declines more rapidly than does the book value of the equipment, it is safest to carry the limits of liability at the aggregate book value. For leased equipment the liability may be on either the lessor or the lessee depending upon how the lease is written, this should be confirmed. If liability is on the lessee then the lessor should be able to provide a schedule of replacement value.

Many policies include exclusion from common perils

which must be removed. These types of peril include electrical service problems; extremes in temperature and environmental changes; flooding; water leakage and overflow; natural disasters such as earthquakes, hurricanes, tornadoes and lightning; civil disorder and vandalism; nuclear damage or radiation.

Another problem which can arise immediately after a disaster can be caused by over-enthusiastic staff attemping to salvage equipment before the assessors have carried out inspection. It is important that clauses invalidating claims should equipment be touched before such inspection are kept to the minimum and that damage evaluation and salvage staff are aware of those which exist.

Automatic coverage for newly acquired equipment is desirable and a procedure should exist to update the fixed asset listing whenever a new piece of equipment is bought and to review the completeness of the list at regular, fixed, intervals.

2. Media

It is not sufficient only to insure for the replacement cost of the media material. The cost of reproducing the information must also be covered and this should be the cost of production from source documents and not just from earlier generation or duplicate media. Should off-site archival or transaction media be destroyed or lost then reproduction will have to be done using source documents.

Again, there are some common exclusions often found which must be deleted and these include coverage for media stored off premises (back-up media must be stored off-site), coverage for media while in transit and mysterious disappearances (media may be stolen and theft be very difficult to prove). Perils commonly excluded for which coverage should be arranged are those listed above in the equipment paragraph plus relative humidity excesses not associated with air conditioning system failures and damage or destruction by electrical or magnetic means.

Although, strictly speaking, valuable papers are not media many data processing operations have a considerable volume of such items as cheques and other negotiable instruments in storage while awaiting processing. These should also be insured.

3. Extra Expenses

Surviving and recovering from a disaster involves expenses which are unique to those activities and, therefore, not included in normal operating budgets. These extra expenses can often be covered by insurance.

They include the cost of using service bureaux; the charges of alternate site facilities; the cost of overtime and extra mailing; travel and hotel expenses.

Other costs which should be considered are overlap in fees charged for use of a hotsite and a coldsite during the preparation to move from one to the other. This might take two to three weeks. Another is the overlap in lease and/or rental payments on equipment when preparing to move from a coldsite back to the permanent data centre.

Some policies are written with a specified time limit on extra expense, such as 30 or 60 days. A worst case recovery effort could take as long as nine months to a year. This time period can be defined, for example, as "such period of time as would be required through the exercise of due diligence and dispatch to rebuild, repair or replace . . .", with no upper time limit and that wording is recommended.

One common exclusion in policies for extra expense which should be avoided is for power failure originating more than one hundred metres from the covered premises. Should there be a general power failure in the area for a sufficient time to cause the expense of activating the survival plan then this extra expense would not be covered.

4. Business Interruption

During a disaster, there will probably be an interruption of processing which may last from 24 hours, if the alternate location is a hotsite, to as long as 14 days if it is a coldsite. Coverage should be arranged to protect against lost revenues due to such an interruption of processing. This might include foregone fees, interest payments and cost of an increased cash float. If an application impact analysis

CHAPTER FIVE

Survival Planning — Basic Considerations

Prepare for the Worst

Survival and recovery from a disaster requires a plan of action designed to minimize the disruption of key business operations. In preparing the plan it is important to realize that disaster and the subsequent recovery is unlikely to be a brief event. Organizations must be prepared to operate in a recovery mode for anything from four weeks to six or seven months depending on the duration of the interruption and the extent of the damage.

Before we discuss the basic considerations for the successful development of the necessary plan it may be worthwhile to consider in a little more detail what such a plan is. Developing a survival plan is in essence performing three tasks. These are first forcing as many as possible of the decisions which must be taken in order to survive to be made before the disaster strikes; second, producing very detailed procedures for all of the hundreds of jobs which must be done during the survival activity, many of which are unique to survival; third, detailing all of the resources which will be needed to do these jobs and making sure that these will be available and ready for use.

There are two main reasons why the procedures developed must be very detailed and why early decision making is necessary. The first is that it is possible that key employees may be absent or injured during the disaster and that their place may have to be taken by alternates who, although trained, will not be as adept as the primary personnel. The second is that the people who are involved in a disaster suffer from a degree of shock and are often incapable for as long as 48 hours after the incident to operate other than as robots. They are incapable of rational thought or decision making or any activity other than following a detailed procedure and check list.

Just as many of the jobs which need to be done to overcome disaster are unique to the survival process and not known during normal business life, so too are the resources required to do them. Many of them are reasonably obvious, certainly to people who have been involved in data processing disaster recovery planning, and include such items as back up information — tapes, discs, operating manuals and so on. Some are not obvious but a little thought will reveal them. For example, suppose the in-house stock of supplies necessary for a survival activity, one that must be restarted in a matter of hours, has been destroyed. It is a strong probability that the normal delivery lead time will not be good enough for survival.

Other resources will be difficult to identify for anyone without experience of surviving disaster and they include such things as the items required to make life bearable for the survival coordinating team cooped up together for maybe days on end in the control centre; protective clothing and communications equipment for staff involved in surveying to assess damage and, even, cash for travel.

All of these resources must be stored off-site. Obvious? A well known university had a major fire in their computer centre and lost all of the back-up information because it was kept in safes in the same building. A small company, which also suffered a fire, could not make their insurance claims in time because the policies were in the building which was destroyed and the only manager who knew the details suffered a heart attack during the emergency.

The resources must also be available, all of them, at any time, day or night, on any day of the year. Disasters have a nasty habit of occurring not at a convenient time, say 3 p.m. on a Thursday afternoon, when the stores and banks are open but at 2 a.m. on Christmas morning in the middle of a blinding snowstorm with no public transport and the roads blocked.

The three activities described of decision taking, procedure writing and resource identification must be complete and must be performed economically, quickly and to a proven method.

Complete because surviving disaster is rather like being born in that it is an all or nothing-at-all thing. You cannot be 99% born and you cannot survive 99%. If you think of 99% of the jobs which will be necessary to survive, develop the procedures and identify the resources for them, it is the 1% which has been omitted which will destroy you.

Economical, because there is no point in spending £1000 to protect £100, almost anything is possible if enough money is spent.

Timely, because experience has shown that if a plan is not completed in some eight months then so much time must be spent in maintaining and updating the parts written that completion to schedule and budget becomes increasingly difficult. Development becomes a moving target.

Finally, the plan must be developed using a proven methodology so that it is structured correctly and so that it is modular. The survival plan is different to any other plan met with in business which usually divided into sections designed to be read in sequence and all of which must be read to provide a complete understanding. The survival plan should be structured so that each individual need only know, be trained in, and be able to use in the worst-case, shock, conditions his or her small section.

It must be modular so that it can be structured properly and also because although the plan is designed to cope with a total disaster, the worst-case situation, in many cases the disasters which occur will be minor or intermediate in nature. The plan must be developed to cope with the worst-case situation but in less catastrophic disasters it should only be necessary to activate those modules strictly needed to cope with what has occurred.

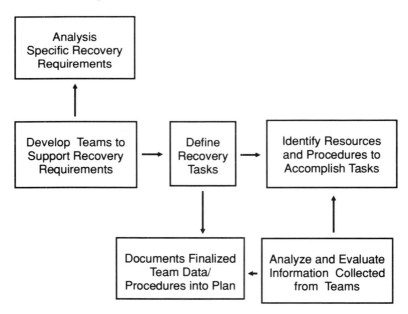

Figure 1: Plan Development

One misconception sometimes met which can be mentioned here is that it is necessary during the survival planning process to know the degree of risk from various possible types of disasters. Although this is necessary to some extent during the prevention activity in order to assist in spending security money wisely, it is not necessary during survival planning. It does not matter what has caused the 'smoking hole in the ground', the necessity is to survive it.

The plan must also be modular so that it can be implemented in easily manageable phases. Usually, although not always, this will start with the central data processing function and the expansion to cover user functions and other vital areas of the organization must be smooth and easily accomplished.

The other basic requirements, common to all survival plan development, which must be met if the plan is to provide adequate ongoing protection are top management support, staff training and good testing and maintenance programmes. In other words, it must be realized from the outset that survival planning is not a once-off activity but is a continuing commitment.

There must be formally stated top management commitment to the process for three reasons: first, because they will have final approval for budget and resource requirements; second, because they will be involved in deciding when to implement the survival plan and at what level; third, because top management are most effective in enlisting the support of all necessary departments throughout the company.

The staff who will have to carry out and manage the survival process must be thoroughly trained so that they can instinctively carry out their part in the plan even in the state of shock immediately following the disaster. In practice this means that it is desirable that the people who will have to use the plan should be those who have been intimately involved in its development. The plans must be comprehensively tested and such testing must be repeated at intervals to ensure that the maintenance procedure is working properly and the plan should contain the procedure by which it will be maintained, kept up-to-date, and designate responsibility for this activity.

Lastly, in this chapter, we should examine the three basic approaches to developing a survival plan. You can adopt a do-it-yourself approach, purchase a methodology book or use specialist consultants.

a. 'Do-It-Yourself'

The do-it-yourself approach requires considerable resources to decide how to develop the plan, what it needs to cover, the way in which it should be structured and the hundreds of tasks which need to be done during survival' many of which could be overlooked by someone inexperienced in such planning. There are also the problems of adequately training the staff in the use of the plan, of practically testing it and of completing it to schedule.

For smaller organizations, with simple plan requirements, a do-it-yourself approach could be most effective. However, do investigate the resourcing required and the economics, carefully.

b. Methodology Book

There are a number of companies selling 'books' which describe a methodology for developing a survival plan. These 'books' often contain the forms to be used to collect the necessary data. Some companies will write the plan from the data collected while some will lease or sell a computer program into which the data is fed in order to produce a plan automatically.

The major advantage to using such an approach is that the external cost, the price to be paid to the outside company, is relatively small. Take care, however, when considering this approach, to make an estimate of the total costs including those which will be incurred internally.

The major disadvantages of this approach appear to be:

* company resource must be made available to study 'the book' and reach a working understanding of the methodology

* since the methodology is, of necessity, generalized, it must be modified and expanded to suit the particular

organization. Expertise in survival is necessary to do this

* it is often difficult to provide sufficient internal resource to adequately project manage the development so that it is both cost effective and completed before maintaining finished sections of the plan becomes a problem

* company personnel have not been involved in the production of the eventual plan but only in providing information for it. They are, therefore, probably not motivated by it since they feel no sense of involvement in it

* since the plan is 'externally' produced, it must be read and understood before people can be trained in its use.

c. Specialist Consultants

There is a growing number of consultants with expertise in survival planning. Many of these are attached to general consulting companies or firms of accountants but there are companies which are dedicated to this type of consultancy as an exclusive activity.

Some of the companies who supply alternate site standby facilities will also supply survival plan consultancy. A possible problem with this is that one of the most important phases at the beginning of the development of a survival plan is, often, the proper assessment of the alternate site requirement and the management decision between various options based on costed information.

The disadvantage of using specialist consultants is the higher external costs involved (although the total cost is often considerably less). The apparent advantages are:

* many of the consultants will project manage the development to ensure both its timely completion and minimum impact on the client staff normal workload

* many of them, also, will do so to a very detailed specification of deliverables

* there is a marriage between the consultants experience of survival planning and the clients knowledge of their business to produce a plan which is unique and tailored to the client requirements

*   the consultant will be responsible for setting up
    adequate training, testing and maintenance
    programmes

Care should be taken in the selection of consultants and
Appendix 3 contains a list of questions, originally produced
for a briefing note issued by the Computer Service
Association in London, which will help you to satisfy
yourself that the consultant you choose is both professional
and experienced.

CHAPTER SIX

Survival Planning — Outline Methodology

In this chapter we will first consider planning the plan development project then the development of the plan and, finally, the testing and maintenance of it.

1. Project Planning

Most consultants experienced in survival plan development agree that the best method is to form the key personnel in the organization into teams which will be responsible both for developing the plan and for activating it and using it when disaster strikes.

This means that the people who will have to manage the survival will, even in the shock situation, thoroughly understand and feel involved in the tool which they are using — the plan — since they were involved in its development.

Even in the simplest plan for a small data processing centre there are 14 areas of activity or functions, which break down into hundreds of individual jobs, which must be addressed during survival and which, therefore, must be included in the plan (these functions are listed below). More complex plans require many more activity areas.

The 14 Basic Areas of Activity for a Simple Survival Plan
(These may be combined).

Management
Administration
Transportation
Replacement Equipment
Technical Support
Application Support
Communications
Alternate Site
Data Preparation
Production Control
Internal User Liaison
Evaluation
Salvage
Personnel/PR/Security

The number of teams will vary with the size of the organization and also with the type of plan being written, but they must be formed to address all of the functions which are identified as necessary. An early activity in the project planning phase is this identification of the teams required and of the people who will be the team leaders.

Before this can be done, however, management must have selected the survival coordinator (and, of course, his or her alternate) and the functions of this individual are so important that they are the subject of a separate chapter; Chapter 7.

The coordinator will call an initial meeting, let us call it the orientation meeting, at which a number of important things will be done. Among others these include the drafting of the policy statement to be issued by top management to publicize their commitment to the plan and the drafting of the objective of the survival Plan. Examples of these are given in Appendix 4.

Another activity at the orientation meeting is to establish the initial team structure and designate team leaders. In addition to the main teams it will also be necessary to form a number of support teams and appoint support coordinators. Appendix 5 contains a description of some example Support Teams.

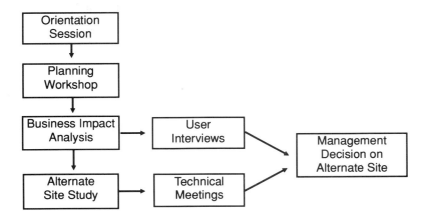

Figure 2: Project Planning

A further activity in the project planning phase is to set the priorities of the activities and business applications. First the survival activities must be identified. These are those which are so critical that if they are not restarted within one or two days then the business simply cannot survive.

Next to be determined are the activities which must be restarted within one or two weeks or they will become survival activities. Finally the criticality of the activities must be assessed by arriving at the financial or operational impact on the organization of their loss. Such impact analysis should always involve the activity or business application owner.

The importance of these setting of priorities and impact analysis activities cannot be overemphasized since they are the foundation of the planning process. The major objective is to provide management with the information necessary for them to make a good business decision about the alternate site. Of course, an important element in such information will be cost, both that of losing the application or business function and that of providing the alternative facility. This is particularly evident when one remembers the '80/20 rule' — 80% of cost comes from 20% of activity. If cost were all that was necessary, however, then the many 'Risk Analysis/Risk Evaluation' computer packages available on the market would be of more practical use than they are. Organizational and political considerations can also be vital in making priority and impact decisions. To take an extreme example, if a government doesn't pay its major creditors then this is probably not vital for a considerable time but if it cannot make unemployment or social welfare payments then this becomes vital very rapidly!

Samples of the types of form which can be used as the basis for gathering the information necessary to do this prioritization and analyses for both business functions and for data processing applications are contained in Appendices 6 and 7 respectively.

It must be assumed that it will necessary to establish an off-site command centre where the team leaders will assemble in the event of a disaster and from where they will control and manage the survival process. At least three such centres must be identified at various distances — say 3 kilometres, 8 kilometres, and 15 kilometres radius — in order to be safe, for example, in an environmental type disaster.

Since the basic guideline in developing the survival plan is 'Plan for the Worst' the plan must always include the ability to move to an alternate location.

In the case of a non data-processing function such planning may be relatively simple, involving such activities as alerting local estate agents of the possible number and size of rooms which may be required and determining where emergency supplies of sufficient furniture, office and communications equipment can be obtained.

For data processing operations however, the evaluation of the various options for an alternate processing capability is considerably more complex. These options include reciprocal agreement, coldsite, vendor agreement, service bureau, hotsite, mobile solutions and redundant site.

The considerations which need to be evaluated, in choosing which option or combination of options will be used, include not only price but also such factors as the organization's tolerance for down time.

If a company can tolerate being without its data processing for seven or more days then a coldsite alone may be perfectly acceptable. As the time required to respond and recover becomes more critical however, additional options become necessary.

Let us examine in more detail the various types of data processing alternate options.

Reciprocal Agreement

This is an agreement between two or more compatible sites that if one site experiences a major interruption in processing then the others will provide back-up facilities during the crisis and recovery period. Economy many appear to be an advantage, there are, however, many drawbacks. As each site progresses it becomes increasingly difficult to ensure that hardware and software configurations remain compatible; both sites could experience the same disaster and it may be difficult to arrange sufficient time for testing. Consultants warn that a probable result of using such an agreement is the transportation of the disaster from the original site to that supposed to be helping because of the unlikelihood of there being sufficient capacity for both organizations' workload.

Coldsite

This option can range from having an empty shell to a pre-engineered facility offering environmental requirements such as power, air conditioning, chilled water, raised flooring, security and communications facilities. When disaster occurs the hardware equipment vendor is called upon for replacement equipment.

Again low cost is a primary advantage particularly if the cost is shared either through subscription to a coldsite standby facility or through a cooperative venture. In addition the coldsite offers economic long-term usage while a datacentre is being reconstructed. Indeed, many consultants recommend this option as the minimum for all survival plans.

There are disadvantages however that often make the coldsite undesirable as the major option. Among these are the long time that it can take to equip and staff the site and make it operational. There is also the possibility that more than one customer sharing the cost of the coldsite may suffer the same disaster. Testing is also a problem.

Vendor Agreement

This is a pre-arranged contract or understanding that the company can use the corporate facilities of a hardware supplier. Considerable care needs to be exercised in concluding such an agreement in ensuring the way in which configurations will be kept compatible; in the arrangements for long term occupancy and in the arrangements in the event of multiple disaster.

Service Bureau

Using a commercial service bureau can have the advantages of immediate availability and the economy of only having to pay for the time used. Long term processing at a service bureau is, however, very expensive and may not be available at all at short notice. Security considerations, testing and ensuring the timely availability of sufficient capacity can also be problems.

HotSite

This option is provided by specialist suppliers of such facilities which offer their clients a fully equipped facility on continuous standby.

In many cases, technical staff are provided as well as communications facilities, data entry, data storage vaults and office facilities. The advantages of this option include long-term availability, security and ease of testing, although it may cost more than other options. One possible disadvantage is that in the event of two subscribers suffering the same disaster, some hotsites operate a 'first come, first served' service. For this reason subscribers to hotsites often also have a coldsite option and indeed many hotsites also offer an adjacent coldsite for the longer term processing requirement.

Mobile Solutions

Both hot and cold options can be provided, under certain conditions, as mobile facilities.

Redundant Site

This is where an organization maintains a totally redundant site in which the datacentre is duplicated at another location. While technically, perhaps, the ideal solution, most companies find it out of the question from a cost view point to maintain identical sites and keep both of them compatible.

2. Developing the Plan

In developing the plan it is necessary to analyse the specific requirements of the organization in terms of what they need to survive.

Many of these are general in nature and common to most organizations, a large number, however, will be unique to the company and function and will depend upon such factors as its structure, its location and environment and the survival strategies which it needs to adopt.

This last consideration, the survival strategies, is vital. It has already been mentioned that although the plan is designed to cope with a total disaster, in many cases the disaster which will occur will be minor or intermediate in nature.

The plan structure should therefore be such that only those modules strictly needed to cope with what has occurred can be activated.

The main purpose of the survival strategies is to define when a disaster is minor, intermediate or major to this company, the escalation mechanism and the overall actions required. These definitions will vary from company to company and will depend, mainly, on their tolerance to down time.

A further step in developing the plan is to define the recovery tasks, the hundreds of different jobs which will have to be performed if survival is to be achieved, and to assign them to teams. Many of these are unique to the survival situation and unknown in the day-to-day work situation. Many are also unique to the company concerned and will depend on its specific survival requirements as established in the first step.

It is necessary to develop the detailed procedures and establish the resources necessary to do these jobs and to document all of these into a plan.

Most professional survival planning consultants agree, however, that these steps — tasks definition, procedure and resource development and plan documentation — are not performed in sequence, with each step completed before the next starts. They should, in fact, be an ongoing loop of activities leading to continuous expansion and refinement.

There are two main reasons for adopting this iterative method of developing the plan. The first is that it assists in establishing planning as a part of the everyday working methodology: people are most likely to realize, when something changes, that the survival plan is affected and must also be brought up-to-date. The second is that it is the basic method of employee training. Involvement in the gradual expansion in the detail of the plan leads to the instinctive understanding of it, which enables its successful implementation in the first 48 hours of shock. Chapter 8 is all about training.

It is very important that, as the plan is gradually developed during the iterative process, good project management tools are in place which ensure that the people involved have a thorough understanding of the work to be done — the overall project picture, what has been completed and what remains. One successful method of doing this is by using charts which at the beginning show the activities for which a team is responsible and which are gradually added to as the project continues to show, from the top down, sub-activities and, finally, survival tasks. Completion of the detailed procedures and the identification of resources can also be indicated on the charts so that the overall amount of work remaining to be done by any team can be quickly determined.

Figures 3 to 6 give an example of how such charts might begin to be developed for the management team. In looking at these it must be remembered that in this form these charts do not represent the time sequence or order in which activities or tasks are to performed but do represent their relationship with each other. The activity chart can be backed up by detail sheets as shown in Figure 7.

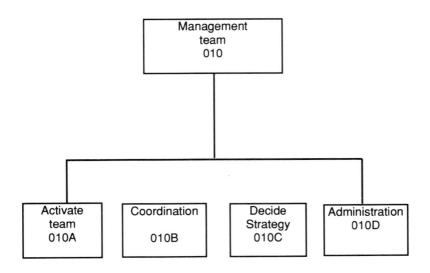

Figure 3: Activity Overview Management Team

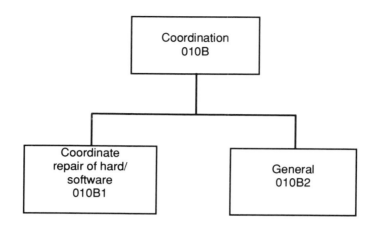

Figure 4: Sub-activities, Management Team Activity B

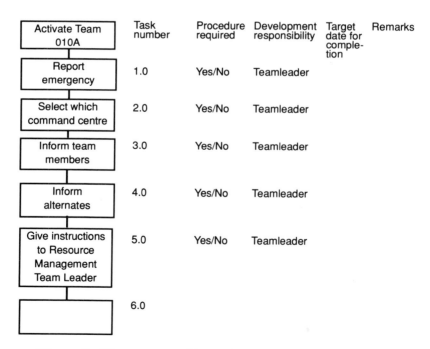

Figure 5: Management Team Survival Tasks Activity A

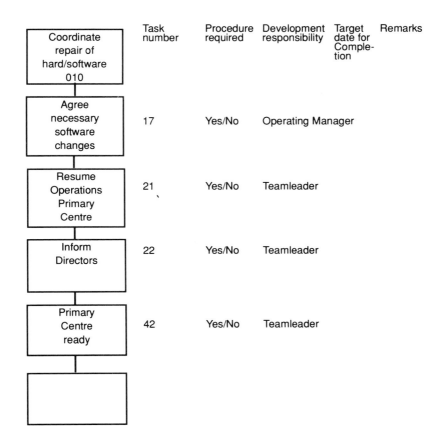

	Task number	Procedure required	Development responsibility	Target date for Completion	Remarks
Coordinate repair of hard/software 010					
Agree necessary software changes	17	Yes/No	Operating Manager		
Resume Operations Primary Centre	21	Yes/No	Teamleader		
Inform Directors	22	Yes/No	Teamleader		
Primary Centre ready	42	Yes/No	Teamleader		

Figure 6: Management Team Survival Tasks Activity B1

Management Function	Alternate Processing Site Function	Data Preparation Function	Production Control Function
Coordinates and monitors all recovery activities	Establishes a processing operations at the alternate site(s)	Preparest input transactions required for production processing	Coordinates production control functions in support of production processing
Activities survival plan and teams	Requests transportation of items from off-site storage	Establishes collection points for input data	Schedules production processing within available time-frames
Determines survival strategy	Schedules computer operations personnel	Schedules data for input to production jobs within available time frames	Coordinates program changes in support of production processing
Recommends either the repair of the primary data centre or the construction of a new data centre	Obtains funding for processing at the alternates site(s)	Maintains data preparation schedules according to the Production Application Priority list	Bursts, decollates and performs any forms handling duties
Coordinates repair and/or reconstruction activities	Assists in security additional hardware	Provides instructions for keying of input	Distributes production output for delivery to users
Monitors required executive support areas			

Figure 7: Examples of Detail Sheets for Some Activities

Finally, during the plan development, the people not included in the current plan should be educated as to their exposure, their needs in protecting themselves, and future plans for expanding the plan to include their function.

3. Testing & Maintenance

The development of a survival plan is not a one-off task which can be considered as complete once the first draft is finished.

There are two other critical components of successful planning; testing and maintenance. In fact, the proper conduct of these two activities is considered to be so important by many of the professional planning consultants that they include in their offered services 'Product Enhancement' contracts which include assistance with and supervision of testing and maintenance.

Chapter 9 is all about testing and so it is sufficient to say here that testing is mandatory to ensure success and is the acid check on whether the plan is workable; of details which may have been overlooked; of timing expectations; of alternate site compatibility and whether backup routines are adequate. There are many different aspects that can and should be tested. Some examples of these are:

* the procedures to notify everybody who needs to know and to activate the organization even at dead of night and, perhaps, during an environmental disaster when normal methods of communication and transportation may also have been lost

* the ability to conduct damage assessment

* how effectively the control centre can be established and how long it takes

* how well applications or, for a non-D.P. plan, business functions, can be brought up at the alternate site.

Integral to the testing process is the maintenance of careful test documentation. There should be a set of objectives to be met for each test and detailed documentation of what was done, by whom, the results, and what the implications are for future testing activity.

Maintenance of the survival plan is also vital. It should

be a living, changing document just as the organization of the company is a living and changing thing. The plan must be continuously updated whenever there is a change in such elements as the hardware, the location of a department or facility, a change in personnel or supplier, the criticality of an application, the introduction of a new system or communications.

Many consultants believe that a survival plan is not complete unless it contains a section detailing the procedure for its maintenance.

Appendix 8 lists the suggested contents for a data processing survival plan. Obviously, for other functions these will be highly variable and will depend not only on the business activity involved but, also, on the structure and complexity of the organization.

CHAPTER SEVEN

The Survival Coordinator

New Roles to Play

The survival coordinator — sometimes called the disaster recovery coordinator — and his or her trained alternates, are the key figures in the survival and recovery process. This is because they play pivotal roles not only during survival plan development but also during normal operations and during the survival and recovery activities.

During the development of the plan the survival coordinator is responsible for:

* reporting to management on the status of the project and spend against budget
* all liaison with external consultants if these are used
* advising on team structure, leaders and members
* arranging and monitoring all meetings
* supervising the preparation and distribution of the plan.

The coordinator has three main roles during normal operations. These are training, testing and maintenance of the plan.

The coordinator also has the responsiblity of advising management when an incident is sufficiently severe to be declared a disaster and then coordinating all survival and recovery activities and reporting on status and progress to management.

The person chosen as the survival coordinator is in a unique position to obtain a very detailed understanding of the organization and operation of the business areas or locations to be included in the plan and such knowledge can be of considerable use in performing non-survival management tasks. It is also true that the level of senior management involvement encountered in the survival planning activity will require high quality written and verbal communication skills. For these reasons it is recommended that, whenever possible, the coordinator chosen is a person considered by senior management to be a 'high flyer'.

The primary focus of the survival coordinator is to maintain a viable and tested disaster recovery plan which demonstrates to management the organization's ability to continue operations following a disruption of services. Maintenance of the plan is ongoing to reflect changes that occur throughout the organization. Testing also occurs on a regular basis to assure an organization wide awareness of the recovery function.

In order to meet these objectives the survival coordinator must:

* identify and review the critical tasks which are essential during a recovery effort

* establish a timetable for regular review and updating of all tasks, resources and procedures outlined in the plan and ensure that updates are incorporated into the plan on a timely basis

* coordinate monthly, quarterly, semi-annual, and annual testing of the plan, report results to management and update the plan as required

* etablish an ongoing training program which ensures employee awareness of the functioning of the plan

* distribute plan materials as appropriate

* establish a standards programme which ensures that changes to critical procedures, functions and documentation are reflected in the plan

* assure that contact is maintained with vendors and support personnel to keep recovery support considerations current

* act as liaison between recovery functions and other organizational areas, including the external and internal auditors, concerning survival planning issues

* meet regularly with survival teams to review responsibilities required during a recovery effort

* maintain contact with local and national emergency organizations which may be involved during a recovery effort

* liaise with organizations which offer survival and recovery services and products

* provide input, support and liaison as required to functional areas which are responsible for maintaining procedures, standards or documents which effect the survival plan. Such things as file rotation procedures, employee telephone list, insurance policies, etc

* research, evaluate, and recommended internal and

external solutions to survival and recovery problems, as required

*   assist divisions in maintaining contracts for alternate facilities and services

*   administer salary, benefit, and appraisal reviews of recovery staff members, if applicable.

The budget for a viable, regularly tested, and maintained survival planning programme must be carefully allocated. A changing organizational environment must be reflected in the survival plan and consideration should be given to the following areas when planning the budget:

*   The monthly subscription cost involved in contracts with alternate site(s)

*   The cost for in-house testing (e.g., personnel, equipment usage, supplies, food, special materials, off-hour access to files stored at an alternate location). The costs for alternate site testing (e.g., transportation of personnel and supplies to the alternate site, food, lodging, equipment usage, employee overtime)

*   The cost for maintaining off-site backup and rotation of critical files and supplies

*   The cost for ongoing training (e.g., supplies, printing, conducting sessions)

*   The cost for printing updates to the plan

*   The cost of providing adequate maintenance

*   The anticipated cost of declaring a disaster which may not be covered by insurance

*   Salary, benefits, and compensation for the survival coordinator position and possible additional staff

*   Possible allotment for important seminars or conferences which pertain to the field of survival planning.

Management makes a substantial commitment of organizational resources to develop and maintain a survival plan. It is therefore essential for the survival coordinator to keep management aware of the status of plan development, testing and maintenance. Such status reporting can be accomplished by first, establishing a yearly planning cycle which outlines training, on-site and off-site testing, and maintenance functions. Second, a monthly report should be prepared detailing the status of active projects and those which have been completed. This report should comment on any problem areas which could prevent or have a severe impact on results.

During the survival operation the survival coordinator will have limited but important duties. He (or she) is the most knowledgeable person in the organization about the plan as a whole and, therefore, his value during the operation will be as an advisor to the management team, of which he is usually a member, and liaison between it and the recovery and support teams.

The choice of which command or control centre is to be used and its activation is the responsibility of the management team during the recovery from an extensive disaster. It is normal that the survival coordinator will be assigned this task.

The location and occupancy procedures of at least three command centres at various distances from the location should have been decided during the development of the survival plan. The Coordinator will need to act as liaison between the team leader in order to establish and resource the command centre at the location chosen as appropriate to the severity of the disaster. The main jobs necessary to do this include:

* contacting the selected command centre to alert them of activation of the site
* retrieving resources from off-site storage
* Alerting appropriate support teams and support coordinators for need of their services such as furnishings, communications, equipment and suppliers
* Providing a supply of logs both for command centre and telephone usage
* Establishing various status charts such as a team status board, an operational function status board, a general message board and, possibly, a personnel accomodations board.

Once the command centre is established, the disaster recovery coordinator will assist in its continued operations.

The survival coordinator is involved in the development and maintenance of all portions of the survival plan. He therefore has knowledge of each element of the plan and the manner in which the plan coordinates the work of all personnel during a recovery effort. This understanding of both the individual elements of the plan and the overall functioning of the plan uniquely qualifies the coordinator to advise management throughout the recovery effort.

Using the survival plan and information from appropriate personnel, the coordinator provides management with clarification of information presented in the plan and most important, the ongoing status of the recovery effort, problems, and decision requirements.

Liaison between the various recovery and support teams is important during the recovery effort, particularly if personnel are physically relocated and therefore normal communication channels are unavailable. The coordinator will be able to answer questions on such items as plan content, responsibilities of various teams and availability of specific support services.

The coordinator will supervise the administration activities such as status and problem reporting and follow up activities when inter team cooperation is needed.

This, necessarily brief, overview of the duties and responsibilities of the survival coordinator will, I hope, underline the necessity of choosing carefully the person to fill this very demanding role and, of course, those who will understudy it.

CHAPTER EIGHT

Training

In a State of Shock

As has been said elsewhere, most of the staff involved in a disaster go into a state of severe shock for some two or three days after is happens. These are, of course, vital days in the survival process when many of the most important jobs with far reaching consequences have to be done. The problem is that in this state of shock people lose their ability to think logically and their memory of how to do things. They can follow a procedure and fill in a checklist but the following day they have to refer back to that list to find out what they did, with whom and how successfully.

It is for this reason, of course, that the procedures in the plan have to be so detailed and the plan itself is so structured that each individual has to know and understand only those two or three pages concerning his or her jobs. That alone is not sufficient, however. Successful survival also depends on training. Training so thorough that individuals in shock, in strange surroundings, sometimes in anguish for injured friends and colleagues, can do their jobs almost by instinct.

Be Prepared

The major training of the original team members and their alternates is the ongoing training which takes place during the development of the plan. Plan development consists of four basic activities once the survival requirements and strategies specific to the organization have been identified. These are defining the hundreds of recovery tasks, identifying the resources and procedures necessary to do these, collecting and evaluating the information necessary and, finally, incorporating it all into the documented plan.

Ongoing team training is achieved by building the plan gradually. First the top level requirements are dealt with and the loop of four activities is performed for these, Then the next, more detailed level is done in the same way — and so on. In this way the detail is built up gradually and is absorbed by the team members rather than having to be learned.

That initial training is, of course, not enough. Team

membership will change, organizational and other changes incorporated during maintenance will have to be learned and training will be required to meet particular circumstances such as a testing program.

It is one of the survival coordinators responsibilities to organize such training and provide the attendees with all of the necessary materials. Experience has shown that this cannot be done in an adhoc manner or only when the coordinator considers it to be required. Training should be formally initiated by operational management submitting to the coordinator a training request form and a sample of such a form is given as Appendix 9.

A training programme is required, therefore, for orienting new team members, reviewing new or updated information with the survival teams, preparing the terms for plan testing and training alternate team leaders and members.

Presenting new or updated information to the teams is often done during a reasonably informal meeting with them and the training required for test preparation is obvious. The test objectives are reviewed, the participation and responsibilities of the teams are discussed and result recording and evaluation procedures are agreed.

The training of new team members and alternate team leaders and members, orientation training, is much more complex and the remainder of this chapter is devoted to describing a possible orientation training programme.

The key points which should be discussed during orientation training and which will be described in detail are:

* Definition of Survival Planning
* Related Elements of a Survival Planning Programme
* Survival Plan Overview
* Recovery TeamsConcept and Overview
* Recovery Strategies, Procedures and Flowcharts
* Survival Plan Maintenance
* Survival Plan Testing

1. Definition of Survival Planning
1.1. Definition of Disaster
    1.1.1.   'Any catastropic or partial disruption to the operational capability'.
    1.1.2.   'A sudden calamitous event bringing great damage, loss or destruction'.
1.2. Survival Planning is
    1.2.1.   Identifying prior to a recovery, all critical procedures and resources necessary to survive the disaster.
    1.2.2.   Developing the most efficient method possible to recover from a disaster.
    1.2.3.   Developing a plan of action to protect the organization's assets, assure organizational recovery, and safeguard employee jobs.
    1.2.4.   Testing the plan of action to ensure that it works.
1.3. Survival Planning is not planning for
    1.3.1.   Changes in business environment such as interest and exchange rate fluctuations.
    1.3.2.   Regulatory changes.
    1.3.3.   Competitive attacks.
    1.3.4.   Extreme technology advances.
1.4. Government regulations which apply.
1.5. Review of Survival Planning Terminology.

2. Related Elements of a Survival Planning Programme
2.1. Organization-wide awareness
    2.1.1.   Protection of organization continuity with a survival plan.
    2.1.2.   Need for survival plans for the total organization.
2.2. Physical security programme
    2.2.1.   Controls and monitors access to sensitive areas.
    2.2.2.   Keeps general public out of organizational areas.
    2.2.3.   Safeguards employees and customers.
2.3. Information security programme

2.3.1. Defines critical files.
2.3.2. Limits access to critical files to those who have a need to know.
2.3.3. Monitors access to all files.
2.3.4. Defines what can be done with files.
2.4. Impact Analysis/Alternate Site Study
    2.4.1. Applied for functions or departments which require specific alternate sites for the re-establishing of operations.
    2.4.2. Cost/Benefit Analysis
* Cost:Expenses associated with establishing, maintaining and utilizing various alternate sites.
* Benefit:Financial and operational impact of not accomplishing the divisional functions or the processing of production applications.
* Management decision:which sites should be used.
2.5. Off-site rotation of critical files
    2.5.1. Defines critical files.
    2.5.2. Assures backup and rotation for currency of critical files to secured off-site storage.

3. Survival Plan Overview
Follow the Master Table of Contents of your plan as a guide for discussion (see Appendix 7).

4. Recovery Teams Concept and Overview
4.1. Importance of Teams
    4.1.1. Small work units.
    4.1.2. Key functional personnel.
    4.1.3. Specific recovery tasks assigned for recovery of its function.
4.2. Team Development
    4.2.1. Major Functions or Departments.
* Identified functions required during a recovery effort are assigned to teams organized in a manner parallel to the operational organization.
* Teams address recovery functions

                  which are similar to their day-to-day functions.

4.2.2.   Other departments
* Functions accomplished by other departments are identified and assigned to teams of key personnel who normally perform or coordinate those functions.

4.2.3.   A division management team supervises and coordinates the functional teams within its division.

4.2.4.   Support teams are formed
* To provide needed support services to the recovery teams for their recovery effort.
* These support services are not normally provided by the divisions themselves.

4.2.5.   A leader, members and alternates are assigned to each team.

4.2.6.   Tasks, resources and procedures are documented in each team's subsection.

4.2.7.   Review of a synopsis of all teams.

4.3. Recovery Strategies

4.3.1.   General Overview: Outlines plans of action based on extent of damage to the primary facilities; use 'Recovery trategies' documented for each division management team subsection.

4.3.2.   Recovery Procedure: Details sequence of tasks requird for each recovery strategy. There should be a flowchart corresponding to the recovery procedure.

5. Example of Sequence of Survival Process

5.1. Security or facility management personnel are informed of an emergency.

5.2. Security or facility management notifies appropriate emergency service (fire, police, etc.) and the Survival Coordinator.

5.3. Survival Coordinator notifies the survival control team leader.

5.4. Survival control team leader notifies the survival control team and the evaluation team leader and establishes the command centre.

5.5. Evaluation team leader notifies his team and eports to the command centre.

5.6. Survival control team leader notifies the affected divisions management team leaders.

5.7. Division management team leaders notify their teams and report to the command centre.

5.8. Evaluation team is dispatched to the disaster site when cleared.

5.9. Evaluation team presents an initial damage assessment report to the survival control team and affected division management teams.

5.10. Executive management is informed of the situation.

5.11. Executive management on the recommendation of the division management teams, determines whether or not to declare a disaster.

5.12. Division management teams notify the functional team leaders.

5.13. Survival control team notifies the appropriate support teams.

5.14. Functional team leaders notify their team members.

5.15. The functional team members notify their staff.

5.16. Teams work to restore their functions at alternate sites.

6. Training

6.1. Orientation Training: for personnel new to or transferring into a department included in the el from other areas.

6.2. Team Training: for team leaders, members and alternates to assure knowledge of functions during a recovery effort.

7. Survival Plan Maintenance

7.1. Purpose

7.1.1.   Effectiveness of the plan.
7.1.2.   Plan should reflect changes in the organization.
7.2. Types of Maintenance
    7.2.1.   Coordinated by survival coordinator.
    7.2.2.   Follows the maintenance matrix.
    7.2.3.   Includes review of changes submitted by a change request form.
    7.2.4.   Conducted during quarterly visits; various elements examined.
    7.2.5.   Annual planning conference.
    7.2.6.   Annual critical file identification and rotation review.
    7.2.7.   Annual training and testing review.

8. Survival Plan Testing

8.1. Purpose
    8.1.1.   Validate the efficiency and effectiveness of the plan by using it in test scenarios and following documented procedures.
    8.1.2.   Validate team memberships and assigned tasks.
    8.1.3.   Evaluate the accuracy of information contained in the plan.
    8.1.4.   Enhance the plan by identifying problem areas or areas that need improvement.
    8.1.5.   Satisfy requirements of executive management, auditing and regulations.
8.2. Methodology
    8.2.1.   Identify the area to be tested and type of testing to be used.
* Component Testing: Focuses on individual elements, functions or sections of the plan.
* Integrated Testing: Combines various components to create a test of related elements.
* Recovery Drill Testing: Encompasses the entire plan; is attempted only after many component and integrated tests.

8.2.2.   Establish objectives of the test.

8.2.3.   Document the findings and results of the test, and distribute to appropriate areas.

8.2.4.   Incorporate needed changes to the plan as indicated by the findings and results of the test.

8.2.5.   NO TEST IS A FAILURE!

CHAPTER NINE

Testing the Plan

Get People Involved

Regularly scheduled survival plan tests must be conducted to validate the efficiency of the plan, evaluate the accuracy of the information it contains, provide the survival teams with the opportunity to interact during a survival scenario, and to complement maintenance by highlighting areas of the plan requiring further development or improvement.

The foundation of a successful testing programme is the participation of employees and management who are confident and knowledgeable of the survival plan. A viable plan is one that is maintained and tested regularly. To be successful and support this objective, the testing programme must begin simply and build on a solid foundation, escalating gradually.

This chapter describes a progressive testing methodology, which moves from testing individual portions of the plan to a comprehensive disaster drill. The methodology presented in this chapter is divided into three broad categories.

1. Component Testing: The first phase of building a successful testing programme is to test the individual components which comprise the plan. Discussed here is the necessity for conducting tests on the parts prior to testing the whole plan, as well as the need to continue component testing as part of a dynamic testing programme.

2. Integrated Testing: This second phase of the testing programme involves combining any number of the components in the order that would occur during survival operations. The methodology of beginning simply and progressing to increasingly complex tests is discussed.

3. Disaster Drill: This last phase of the testing programme briefly discusses activating the entire survival plan.

This approach is illustrated by the schematic given in Figure 8 in which the module selected at the beginning of the sequence can be a component of the plan, a combination of a number of components or the total plan. In other words, these modules are the activities or tasks developed in the charts shown in Figures 3 through 6.

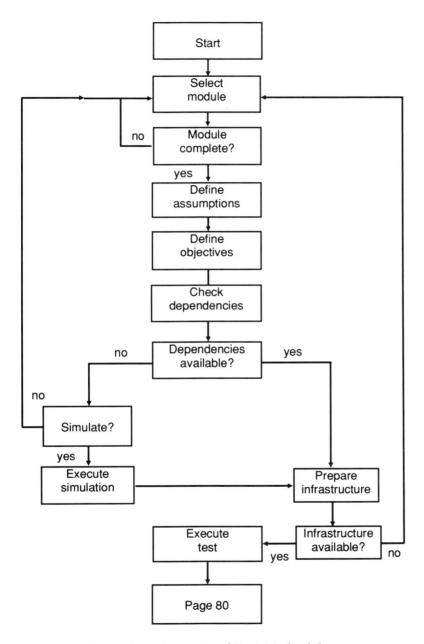

Figure 8: Schematic of Test Methodology

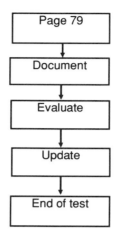

Figure 8 continued

Application of this systematic approach to the development of a testing programme will promote a greater understanding of the functioning of the plan for all personnel associated with it. Because the testing process is a learning process, NO TEST IS A FAILURE.

Component Testing

As the first phase of developing a systematic and consistent testing programme, component testing focuses on those parts of the survival plan which can be tested individually. By allowing personnel to familiarize themselves with portions of the plan through component testing, a greater understanding of the plan's functioning can be realized. Component testing will also prepare personnel for more complex integrated testing.

Component testing is often conducted informally or in a 'walk-through' scenario, even so objectives must be established. Procedures are examined, such as those for team tasks, transportation of personnel and supplies, or off-site storage of critical files and materials. Information contained on checklists, telephone lists, inventories, vendor lists and other support resources is verified for accuracy.

Since many of the components change quite often, testing of each component occurs on a more frequent basis. Even though component testing should be conducted as a first phase to establishing a testing programme, it should also continue once successful integrated testing is achieved. Due to the high visibility of the project, this frequent testing will also generate greater organization wide awareness of the survival plan.

The coordinator will schedule component testing and these guidelines can be followed for all component tests.

* Notify individual or team of test date listed on test schedule at least two weeks in advance; ensure that the test location can accomodate personnel.

* Have test personnel bring their portions of the plan if they have been given a copy.

* Be prepared to distribute copies of plan material being tested, in the event personnel do not have copies.

* Document one or two objectives for the material being tested, such as:
  a. Verify that listed telephone numbers are correct.
  b. Determine if identified resources are correct.

* Discuss objectives with test personnel.

  DO NOT DEVIATE FROM OBJECTIVES DURING THE TEST.

* Conduct the test using status reports, logs, and checklists as a means of documenting results.

* Publish a test report to management involved soon after the test.

* Update the plan as necessary.

The following test synopses are suggested for component testing.

1. Individual Team Test: Following the testing schedule, the coordinator notifies the test team of the upcoming test session. These guidelines can be used:
   a. Call together primary and/or alternate team members.
   b. Distribute a prepared test scenario which realistically represents a situation which could call for activation of the survival plan.
   c. Allow team members to follow their team subsection

to address the disaster scenario in a walk-through manner.
   d. Encourage use of any pertinent logs and checklists.
   e. Document telephone calls made.
   f. Instruct team members to check accuracy and availability of resources and requirements.
   g. Limit the test to approximately one hour.
   h. Document and publish results.
   i. Update the plan as necessary.
2. Establishing a command centre: The coordinator along with the survival control team should follow procedures in the last part of this chapter to examine command centre locations. The following should be considered:
   a. Are command centre locations current?
   b. Are recommended requirements available at the site designated? Or, are procedures in place identifying how to acquire the materials?
   c. Are chairs and tables available?
   d. Are telephones available?
3. Notification Test: Each team will have a copy of the section containing a calling procedure and telephone log topic. A notification test can be conducted, by team, following these guidelines:
   a. Select a division for testing.
   b. Request the team leader or two team members of the management team, to conduct the test and contact functional team leaders.
   c. Instruct functional team leaders to use the calling procedure and telephone log and notify team members and personnel identified in their team subsection.
Management team members can be asked to call required support coordinators.
   d. Have callers state the nature of the call (a test) and verify office and home telephone numbers.
   e. Establish time parameters within to conduct the test.
   f. Document and publish the test results.
   g. Update the plan as necessary.
4. Off-site Storage Requirements: The coordinator should conduct this test with the team members who are assigned to monitor off-site stored resources. There are

some differences, in types of resources which are stored off-site and its documentation, between data processing and other divisions. Those differences are indicated below.

a. Using the off-site storage checklist, confirm the validity of resources which are supposed to be stored off-site.

b. Using the checklist, go to the off-site storage location and verify that the identified items are actually stored off-site.

Additionally for Data Processing:

1. Using the recovery file requirements listing and file restoration list, confirm that critical files are being copied for off-site storage.

2. Go to off-site storage location and verify that identified critical tape files are in fact off-site and being rotated properly.

Additionally for other divisions:

1. Using off-site storage checklists, again, confirm the validity of critical hard copy or magnetic media files which are supposed to be rotated off-site.

2. Go to off-site storage and verify that the appropriate critical files are in fact stored off-site.

c. Document discrepancies

d. Establish work plans to improve the off-site of resource and critical files.

e. Publish the results of the test to appropriate management.

f. Update the plan as necessary.

5. Availability of Supplies, Equipment and Services from External Sources: The coordinator will schedule a vendor and servicer notification test with selected teams to verify that the required support resources are properly identified and available.

a. While regular maintenance procedures should keep the documentation current, a telephone test to the various vendors should be conducted to validate and correct not only telephone numbers and contract information, but also supplies and services offered by them.

b. If letters of intent have been received from vendors,

review the commitment stated in the letter for accurancy and feasibility.

c. If letters of intent are applicable but will not be submitted, request verbal confirmation of support from the vendors in the event that they are called upon for assistance.

6. Availability and Feasibility of Requirements to be Provided by Support Teams: The coordinator will review with each support team the checklists which are applicable. The review should include the following points:

    a. Total quantity of critical supplies available in current warehouse inventory or adequately available through reorder process established.
    b. Missing information.
    c. Unclear information.
    d. Erroneous information.
    e. Inappropriate items on checklists.

    Document any discrepancies and report these to the appropriate team leader. Obtain corrections or make adjustments and update the documentation. If inadequate support is available, formulate a work plan to identify adequate sources or adjust requirements.

7. Alternate Site Test: If a specific alternate site has been designated, numerous tests will be required.

    The names, telephone numbers and other contact information for both primary and alternate contacts must be verified. Documentation of the agreement allowing use of the site and its significant features should exist and be reviewed for adequacy.

    Either through a personal visit or that of qualified representatives, the site must be reviewed to assure that it is still usable and accessible in an emergency. Make sure that the time allotted for occupying the site and the agreed provision of resources (i.e., equipment and supplies) are sufficient to meet emergency requirements.

    If specific equipment is involved, a variety of tests will be required. This will assure continued compatibility with equipment being backed up and with other equipment with which it will have to interface. For instance, for data processing the following component testing may be used.

   a. Start up alternate site CPU and load the operating system.

   b. Start up and load operating and application systems. Process some batch applications.

   c. Start up and load operating and application systems. Load telecommunication system. Process some batch and on-line applications.

8. Alternate Services Test: The first test to be performed is to verify whether or not the contact information is accurate. The accuracy of the name(s), telephone number(s), and other needed information must be assured.

Any agreement with the servicer describing the services to be provided, the cost and any other conditions or specifications related to the delivery of services should be documented. The documentation of this agreement must be reviewed to assure that it is current, and it appears to have been signed by an authorized representative of the supplier.

Due to changes in equipment or procedures, materials or data supplied to or received from this servicer may not be compatible. This should be tested to verify that compatibility still exists.

The time required for this servicer to perform his function, together with the time required to provide input and receive the output, should also be estimated and evaluated. The key question is whether these services can be provided in the timeframe required during an emergency.

Integrated Testing

The second phase of the testing programme involves integrating any number of the components in the order that they would occur during actual recovery operations. Integrated testing builds on the successes and employee awareness of survival planning generated during component testing. While integrated testing can begin prior to completing all component tests, the coordinator should be cautioned to follow the guidelines below in beginning an integrated testing program.

The increased complexity, involvement of multiple teams and other interested personnel, as well as budget considerations, limit the frequency of integrated testing. However, the benefits of well-planned integrated tests are considerable, even if they are only conducted annually.

When planning an integrated test, the coordinator should follow these guidelines:

1. Integrated testing offers the coordinator the means with which to build organizational confidence in the workability of the plan. As outlined in the sample tests below, integrated testing should begin simply and increase in complexity with each successive test. This approach is recommended in order to familiarize personnel with the plan and better prepare them for a recovery operation. Highly complex integrated testing is discouraged in the initial stages.

2. Audit involvement in integrated testing is strongly encouraged. Auditing may assist in establishing objectives and determining what should be tested from an organizational perspective. By encouraging auditing to issue an endorsement of the test objectives as well as a post-test report of testing procedures and results, organization wide awareness of the plan will be enhanced.

3. Test success can be enhanced by conducting pre-test planning meetings with teams and interested personnel. This is especially true when test schedules are being established. Input from other areas which may be impacted by personnel being absent from their normal duties will be beneficial. Initial integrated testing should be scheduled to minimize the impact on operations.

4. Test objectives should be determined well in advance of a test. They should be concise and realistic. Objectives should not be too numerous as this may increase the possibility of failure to complete them all thereby discrediting the test to some personnel. Test times, including breaks, should also be considered and included with the objectives. Two sets of sample objectives in a format which can be used by the

coordinator during the test for evaluation of each objective are given on page 72.

Information from this form can then be used to document test results. It should be stressed during the test that objectives will not be compromised. Compromising the objectives will only invalidate the test.

5. The coordinator should instruct participants that the status reports and progress logs should always be completed during integrated tests. During testing, enhancements to the plan may become apparent.

Status reports and progress logs can provide the coordinator with information which can be used to enhance the plan. The actual time elapsed for each task can be documented in chronological order. During an actual survival operation, management can use time estimates in planning recovery strategies.

6. Creating realistic recovery scenarios for distribution during testing activities can be beneficial in setting the tone of the test. As outlined briefly on page 73, two kinds of scenarios should be presented to test teams and interested personnel:

   a. A general scenario which briefly describes the disaster being responded to. Teams will be able to respond to the scenario when determining what course of action to take.

   b. Specific team scenarios should be developed for the various teams being tested to encourage greater thought and interaction.

   Scenarios can be developed which focus on the geographical area's potential for specific types of natural disasters, fires, or sabotage. They may also require that teams must go outside the organization to gather information needed in order to complete the test. Effective use of scenarios can also generate discussion on aspects of the plan, such as procedures, which may require enhancement.

7. In planning integrated tests which require travel to an alternate site, the coordinator must consider the following budgetary items (NOTE: this does not represent a complete list):

    a. Transportation costs for personnel and materials to and from the alternate site.

    b. Insurance costs for shipping materials.

    c. Lodging, food and transportation of personnel while at the alternate site.

    d. Needed supplies.

    e. Associated test fees at the alternate site.

    f. Overtime wages for test personnel.

While many of these items may be addressed prior to the actual test, if any vendor or department providing these services would normally be contacted by a team member, the team member should contact the vendor or department to verify the service being provided.

8. Input from personnel involved in the test should be sought by the coordinator after the test. The perceptions of those people who use the plan during simulated recovery operations may offer the coordinator insight for further enhancement to the plan. Appendix 10 is a sample questionnaire which should be distributed to test personnel prior to the end of the test. Every effort should be made to gather the questionnaires from test personnel at the end of the test.

As discussed earlier, progressively complex tests should be developed, building upon the foundation established in component testing. Since integrated testing allows for combining any number of sequential components, it is not possible to list all various components which can be created. Listed below, are tests which utilize several components:

1. Walk-through integrated test: Very often the first test conducted after plan development is a walk-through of selected components. The sequence of events to follow in creating a test include:

    A. Select portions of the plan to test. These could be:

       * establishing a command centre
       * notification, using the calling procedure in the common team subsection
       * arranging transportation to the alternate site(s) (checking the availability of transportation)
       * completing the progress logs

*   selecting a recovery strategy
*   having teams contact vendors identified on checklists
*   completing status reports

B.   Establish test objectives; gain audit approval.

C.   Create disaster scenarios which are relevant to the organization; keep them simple.

D.   Arrange for location of one or multiple command centres; assure that telephones are available.

E.   Provide copies of logs, checklists, scenarios, and objectives to be distributed during the test.

F.   Determine the number and names of test personnel (usually all team members or their alternates).

G.   To minimize impact of the test on the working environment — given that all team members within a division are requested to attend — schedule the test for four hours.

H.   On test day, distribute scenarios and allow teams to interact to address the scenario. Individual teams should be grouped together.

I.   Make as many telephone calls to vendors, servicers, travel agents, alternate site, etc., as is possible.

J.   After the test, distribute the questionnaire.

K.   Publish the test results.

L.   Update the plan as necessary.

NOTE: It should be stressed prior to the test that this excercise is being conducted to allow teams the opportunity to interact in using the plan to solve a problem; no one is being 'graded'.

2.   Data Processing Alternate Site Tests: Obviously, the test required by data processing must be comprehensive to ensure effective use of their alternate site. However, due to the complexity of these tests, they must be accomplished step by step, building on component tests. The following are suggested steps in that process.

A.   Follow Up Alternate Site Test: For the followup test at an alternate site, the coordinator directs test participants to concentrate on loading systems software and processing a few batch applications. To aid in achieving test success, it is essential that a hot site machine startup test has been completed. This alternate site test should involve a limited number of teams as outlined below:

1. Select portions of the plan to test. These may include:
    a. Procedures to pull tapes from off-site storage.
    b. Procedures for establishing the command centre.
    c. Procedures to gain access to the alternate site.
    d. Procedures to load systems software.
    e. Procedures to load and process production applications.
    f. rocedures to distribute output.
2 Establish test objectives based on those portions of the plan being tested; gain audit approval.
3 Arrange test time at the alternate site.
4. Conduct pre-test training meetings.
    a. Emphasize that this test is intended to familiarize personnel with the alternate site.
    b. Select a few key applications to test.
    c. Select team members who will travel to the alternate site; other members remain at the command centre where they monitor test activities.
    d. Inform test participants of test objectives.
5. Assure that logs, checklists, and scenarios are available for distribution during the test.
6. Provide transportation for personnel and supplies travelling to the alternate site.
7. Conduct the test at the alternate site.
8. Balance reports generated during the test.
9. Following the test, distribute the post test questionnaire provided in this subsection.
10. Conduct test debriefing with all team members.
11. Publish test results.

12. Update the plan as necessary.

B. Expanded Alternate Site Test: Based on the success of the first two tests which were designed to familiarize participants with the plan and the alternate site, expanded alternate site testing in now possible.

The following guidelines apply:

1. Combining elements of the two prior tests, plan a initial test involving all teams.

2. Expand the number of production applications to be brought up and processed at the alternate site to the maximum realistic level.

3. Include in the test the establishment of the on-line network.

4. Create scenarios which involve processing at the alternate site. Personnel going to the alternate site are selected during the test.

5. Establish objectives, schedule the test and arrange transportation for personnel and supplies travelling to the alternate site.

6. Conduct the test by first testing all components which can be tested at the command centre and then proceeding with testing of components at the alternate site.

7. Balance reports generated during the test.

8. Following the test, distribute the post test questionnaire provided in this section.

9. Conduct test debriefing with all team members.

10. Publish test results.

11. Update the plan as necessary.

3. Use of Alternate Personnel: Conduct the test described above in much the same manner — except specify that primary team leaders and primary team members are not available. Alternate team leaders and alternate team members must fill the appropriate positions on the team and carry out the test.

All divisions can expand these tests to included:

    * Coordination with support teams and support coordinator.

* Coordination with other divisions which have a functional impact.

Data processing can expand the testing to include:

* Processing of all priority one and priority two applications.
* Full online processing.

After conducting this series of tests, the coordinator will have exposed a large number of people to the survival process, firmly establishing the credibility of the plan. The plan will have been reviewed and enhanced several times contributing greatly to the accuracy and completeness of it. However, it should be stressed that both component and integrated testing should continue. As operations, functions and personnel change, so must the plan, including new procedures and new tests.

Unannounced testing is now recommended. Disaster drill testing more closely simulates actual post-disaster conditions.

Disaster Drill Testing

This last phase of the testing programme should only be attempted after extensive component and integrated testing has been completed. The object of such an exercise is to test the team member's ability to use the plan to respond to an emergency.

These points should be addressed prior to the test:

1. Assure that only a few people have knowledge of the test, such as selected senior executives and auditing personnel.
2. Using guidelines in the previous topic, establish measurable objectives.
3. Again using guidelines in the previous topic, create extensive, realistic scenarios which will support the objectives.
4. Select a test date that will not greatly impact normal operations unless testing during high productivity is an objective.
5. Assure that funding to accomplish the objectives is available prior to test activities.

Again, disaster drill testing is designed to test the team

member's ability to use the skills they have practiced during component and integrated testing.

SAMPLE OBJECTIVES FOR A
WALK-THROUGH TEST

OBJECTIVE	COMMENTS

Familiarize team members with the content and organization of information contained in the plan.

Evaluate the clarity of the plan in defining how teams are to interact with one another.

Evaluate the clarity and accuracy of team tasks.

Assure that selected materials and checklists identified in the plan are accurate.

SAMPLE OBJECTIVES FOR AN
ALTERNATE SITE TEST

OBJECTIVE	COMMENTS

Determine if supplies provided by the alternate site are sufficient.

Evaluate the effectiveness and capability of the alternate site as an operational facility.

Test the effectiveness of the off-site storage programme (i.e., determine if the correct files are stored off-site and that hey are available for the test).

Test team-identified procedures for operating at the alternate site.

Assure the test results balance with actual production results.

SAMPLE GENERAL DISASTER SCENARIO

It is 6 a.m. Monday. A fire has occurred at the headquarters building. You have been instructed to report to the command centre selected by the survival control team. The initial report of the evaluation team as to the nature and extent of the fire will be delivered shortly. You have heard, however, that the fire extinguishing system did not function prior to the arrival of the fire brigade.

Until the impacted management team has selected the appropriate recovery strategy, please review these test rules:

a. Accept stated information as fact when scenarios are received.
b. Work with your team and other teams to respond to this scenario.
c. Complete all logs as appropriate.
d. As in a recovery situation, the impacted management team will call an end to the recovery operations for its division.

SAMPLE SPECIFIC TEAMS SCENARIOS
EVALUATION TEAM: Report the following to the management team(s):
a. The fire was deliberately set.
b. Since the fire extinguishing system malfunctioned, extensive damage has occurred on the ground floor.
c. Smoke damage is also extensive.
d. The eletrical facilities are inoperable but the extent of the damage is unkown.
e. Water damage has been noted on all floors.
f. Structural damage to the building may have occurred.
TELECOMMUNICATIONS TEAM: Report the following to the management:
a. Due to a PTT strike, establishment of communications lines to the alternate site(s) may be delayed.
FACILITIES/PROCUREMENT TEAM: Report the following to all teams:
a. Due to a paper shortage, forms vendors will not be able to meet replacement orders for at least one week.

CHAPTER TEN

Auditing the Plan

Expecting Change

And so you now have a plan which is being maintained and tested to a defined schedule. I am afraid, however, that this is not the end because, like any other activity vital to the continuance of the business, you cannot be satisfied unless the plan is periodically audited. Indeed the importance of such audits is recognized in some countries by governments who have mandated that in key industries they are a legal requirement. They have also mandated that the audit must be conducted by an independent third party!

A plan audit must, of course, be conducted against a set of standards and is accomplished by the use of detailed checklists which examine the inclusion in the plan, and the adequacy of, the following:

The Introduction
Corporate Considerations
Functional Priorities and Survival Strategies
Plan Activation and Notification
Survival Teams
Support Teams
Alternate Sites
Alternate Services
Off-Site Storage Procedures
Survival Resources
Training and Testing
Maintenance

An example of typical checklists is given in Appendix 11.

The details which should be reviewed during an audit include:

1. The survival plan objectives.
2. The sequence of restoration for critical functions, application systems and data processing services.
3. Defined function and application priorities and processing requirements with the operational departments.
4. The methodology for activation and management of the recovery effort.
5. Any unique equipment and/or services which should be addressed separately in the plan.

6. Existing retention requirements including any retention, rotation or retrieval procedures.
7. Survival plan notification procedures
8. Examination and evaluation of the adequancy of items stored off-site for survival purposes.
9. Procedures for moving to an alternate operating site.
10. Alternate site support agreements.
11. Alternate services support agreements and specifications.
12. Existing critical procedures for operations identified in the survival plan.
13. Procedures for updating the survival plan.
14. Survival plan testing procedures.

In addition to the above the following points which are specific to data processing should be reviewed:

* data processing functions for survival

* examination and evaluation of considerations for hardware and software to support the survival effort

* alternate site (hotsite/coldsite) support agreements

* documentation of existing configurations and procedures

* documentation of configurations and procedures for the alternate site

Finally, to help you to appreciate how important and how helpful a well conducted audit can, the Conclusions and Recommendations and Summary Matrix Sections of an actual audit report are quoted in Appendix 12 — suitably edited to disguise the client, of course.

This client had spent considerable time and resource on developing a plan some years before the audit was conducted. Since that time they had experienced considerable growth and many organizational changes and the maintenance of the plan had been, at best, spasmodic. Appendix 12 was the result!

CONTINUITY PLANNING

APPENDIX 1

GLOSSARY OF TERMS

(a) Contingency Plan: A plan of action to be followed in the event of a disaster or emergency occurring which threatens to disrupt or destroy the continuity of a normal business or service provided by the organization, together with the necessary procedures and definitions of required resources. Continuity plans, survival plans and recovery plans are examples of contingency plans.

(b) Continuity Plan: A full continuity plan consists of a security review necessary to minimize the risk of disaster occurring and the survival plan necessary to ensure recovery.

(c) Disaster: Any accidental, natural or man-made malicious event which threatens to or does disrupt normal operations or services for sufficient time to affect significantly, or to cause failure of, the organization.

Accidental disasters include such events as fire or errors and omissions; natural disasters include flooding, blizzards and electrical storms; man-made, malicious disasters include computer related fraud, sabotage, hacking and industrial espionage.

(d) Security Review: A comprehensive, periodic review of all elements of security of both tangible and intangible assets covering policy, effectiveness of policy implementation, restriction of access to the assets, accountability for access and basic safety.

(e) Survival Plan: The plan of action to be followed in the event of disaster in order to be able to survive and recover from it.

(f) Recovery Plan: That part of a survival plan which addresses the activities of restoring the operating efficiency which existed before the disaster.

(g) Alternate Site: Any operational location which substitutes for the primary facilities during a survival effort.

(h) Additional Operating Site: Generally associated with data processing survival, any site other than a hotsite or coldsite; includes facilities used under reciprocal agreements and services bureaus. Functions which support data processing (e.g., data entry, report distribution) may be processed at an additional operations site.

(i) Non-Stop Standby: Parallel processing with two or more integrated processors within a single configuration on the same site. In-house protection, with no standby for the computing environment or communications.

(k) Flying Start Standby: Parallel updating at separate locations with immediate availability of a standby processor. This requires dedicated equipment and is therefore normally an in-house option.

(l) Hot Start Standby: A complete computing environment with an empty machine available within hours. Can be static, transportable or mobile. Availability is limited to a strictly controlled number of subscriber machines.

(m) Cold Start Standby: A permanent or temporary/transportable computing environment, with power supplies into which separately sourced system hardware can be installed within days.

(n) Off-Site Storage for Backup Data: A secure media library with full controls, situated outside the geographic area that might be affected by the disaster. The provision of scheduled daily, weekly or 'on demand' exchange of backup tapes/disks etc., plus round-the-clock emergency support.

(o) Alternate Services: Services which are offered by an external organization for a fee which could replace services normally performed internally in an organization following a disaster. Examples would include microfilm and microfiche production, coin counting and packaging, photocopy services and bulk mailing.

(p) Element Failure Impact Analysis (EFIA): A reference matrix which documents the general impact (down until element is restored, down until manual back up is effected, degraded until element is restored, no effect) on a production application or an operating system should an element of data processing equipment or operating system be rendered inoperable.

(q) Incident Escalation Process: A set of documented steps that trace notification of appropriate personnel following an abnormal or disruptive event originating within the primary data centre; the process includes decision points where a problem becomes a disaster, making activation of the survival plan necessary.

(r) Master File: A file that is either relatively permanent or that is treated as an authority in a job.

(s) Recovery File: Master or transaction files which are necessary to recover and resume processing of a production application system. These should be backed up on magnetic media and stored off-site.

(t) Transaction File: A file containing relatively transient data that is processed in combination with a master file (e.g., a transaction file indicating hours worked processed with a master file containing employee name and rate of pay).

(u) Support Function: A function which would be needed by a department of an organization to effect its survival process, which it does not normally provide within the organization (e.g., physical security, personnel, and transportation) for which a team may be formed.

(v) Support Coordinator: Those support areas which would be needed by a department's survival effort but which are not complex or extensive enough to require a survival team can be provided by single person assignments, known as support coordinators. Examples would be legal services and insurance services.

APPENDIX 2

LIST OF SECURITY REVIEW TOPICS

The following are the major items that are investigated during the security review:

Building Security
Computer Room Security
Human Resources
Computer Operations Security
Network Security
Data Security Administration
Office Automation
Central Security Policy
Internal Audit Controls
Technical Computer Security
Functional Area Security
Distributed Data Processing Security

The topics to be reviewed for each of these major items are detailed in the rest of this appendix.

Building Security

Building Security covers major areas that are of significant value to your disaster recovery planning.

1) Layout/Floor Plan
2) Construction/Building Materials
   a. Review environmental equipment to identify potential problems.
3) Fire Protection/Prevention and Safety
4) Heating and Cooling
5) Electrical
   a. Investigates site construction with regard to water and electrical concerns.
6) Access
   a. Intrusion prevention, detection and response.
   b. Identification of risks due to site location.
7) Services Personnel
   a. Cleaning.
   b. Maintenance.
   c. Outside vendors.
8) Security Staff
9) Terrorist Considerations

Computer Room Security

1) Layout/Floor Plans
2) Construction/Building materials/Water protection
3) Fire Protection/Prevention
4) Environment
   a. Heating/Cooling.
   b. Cleanliness.
5) Electrical
6) Access
7) Service Personnel
   a. Customer/Field Engineers
   b. Utilities
   c. Cleaning/Maintenance

Human Resources

The human resources portion of the study covers a wide range of personnel related procedures, practices and potential problems.
1) Hiring Procedures
   a. Background checks
   b. Orientation
2) Termination Procedures
   a. Voluntary
   b. Involuntary
3) Ongoing Personnel Practices
   a. Grievance procedures
   b. Security awareness training
   c. Internal communications
   d. Performance reviews
   e. Vacation/Sick allotment time
4) Contract Staff

Operations Security

Operations Security covers six main areas in detail:
1) Policy and Procedures
2) Production Control
   a. Run Balancing
   b. Scheduling

    c.   Change Control
    d.   Quality Assurance
    e.   Equipment Utilization
3) Operator's Functions
    a.   Shift rotation
    b.   Duties
    c.   Procedures
    d.   Run documentation
4) Tape Management
    a.   On-site library
    b.   Off-site storage
    c.   Software/Procedures
5) COM/Disbursements
    a.   Output controls
6) Data Entry
    a.   Data security
    b.   Data back up

Network Security

1. Terminal Security
2. Remote Job Entry
3   Telecommunications, Security (e.g., dial-in lines, lease lines, etc.)
4. Vendor-supplied Timesharing Services
5. Network Control Centre Configuration and Operation
6. External and Internal PBX and Switching Facilities

Data Security Administration

Data Security in an in-depth audit involving:
1) Systems Access Control (Logical Security)
    a.   Software security
    b.   A review of actual system, subsystem, utility, database, library and file access control and integrity
    c.   Operations standards
2) Information Control
    a.   Ownership of information and associated responsibilities

     b. Information on security policies, practices and procedures

     c. Tape management system

3) Programs

     a. Operating system software

     b. Application system software

     c. Packages, both for system management and security

4) Communications

     a. Production on-line monitors (CICS, IMS, IDHS, DMS2, DB2)

     b. Timesharing systems (TSO, CMS, CANDE, TSS, UFO, FOCUS)

     c. Network software (NCP, SNA, SDLC)

Office Automation

Office Automation includes coverage of such topics as:

1) Minicomputers, Microcomputers, PCs
2) Microform

     a. Microfilm

     b. Microfiche

Central Security Policy

1) Descripiton of central policies.
2) Awareness of employees.

Internal Audit Controls

This section reviews the internal auditing procedures for:

1) Building Security
2) Computer Room Security
3) Human Resources
4) Operations Security
5) Network Security
6) Data Security Administration
7) Office Automation

Technical Computer Security

1) Computer Operations
2) Quality Assurance
3) Data Security
4) Data Base Management
5) Technical Services
6) Application Programming
   - development standards
   - maintenance standards
7) Network Security

Functional Area Security

For each relevant functional area all the above mentioned security aspects are discussed with the managers of these areas with special attention to the specific problems related to the use of personal computers.

Distributed Data Processing Security

A separate review of all automated service suppliers and of all areas where independent processing is being conducted must be performed. This review should cover all of the items which are included in the other reviews.

APPENDIX 3

PLANNING CONSULTANCY — BUYER'S QUESTIONNAIRE

Purpose:

The methodology used in developing a plan may, quite legitimately, vary between the various consultancy companies able to assist you. This questionnaire should enable you to satisfy yourself that, no matter which methodology they use, the consultant you choose will;

* have the level of professional experience you require
* give the degree of management support necessary to your organization during plan development
* provide adequate training, testing and updating procedures.

Questionnaire:

1. How will the consultants project manage and schedule the development of the survival plan?
2. How will the consultants ensure that the involvement of your staff is kept to a minimum during the survival plan development (so that normal work and operating efficiency is not impacted) but also ensure that there is sufficient involvement by them properly to identify the survival requirements unique to your location, organization and business?
3. How does their methodology allow for a phased survival plan development; so that a plan initially developed for data processing can be smoothly extended to include other parts of your organization, should this prove desirable?
4. Do the consultants provide an impact analysis detailing the financial and operational impact of losing applications or functions? What methodology is followed in arriving at this assessment?
5. Will the consultants assist in evaluating the effectiveness of your data processing insurance?
6. How are the consultants involved during the training and testing phases?
7. How do the consultants ensure detailed plan maintenance?
8. How will the consultants supply support on site during

a disaster to help your staff implement the survival plan?

9. How many survival plans have the consultants produced and over what time scale?

10. Have any of these plans had to be activated? If so, with what result?

11. Do the consultants conduct a security review in order to minimize the risk of disaster?

12. Will the consultants assist in developing solutions to problems revealed by the security review?

13. Is this activity their total business? If not, what proportion of their business and staff does it represent? How will the consultants ensure that there is no conflict of interest between this activity and other business activities?

14. What is the method of payment, fixed price or time and material? If the latter, how will the consultants ensure completion to budget?

APPENDIX 4

POLICY STATEMENT

It shall be the policy of **(organization name)** to provide a survival plan to protect the assets, accurate records, the well-being and safety of employees and to provide for the continuation of essential services to the organization and to its customers.

The plan will provide for the re-establishment of critical operations within _____ hours of a declared disaster.

The plan will provide for the restoration of these critical operations in the defined priority sequence.

The plan will be the responsibility of **(department name)** to ensure continued maintenance and quarterly review and testing.

Approved by:
(Signature)
(Title)

Objectives of the Survival Plan

The objectives of the survival plan are to provide a programme to achieve the following ends in the event of a disaster in the facilities of **(organization name)** located in **(city, country)**.

Major Objectives

1. To continue operations in order to maintain essential customer services, to continue support services to maintain cash flow and to maintain the confidence of customers, vendors, employees and shareholders.
2. To provide for the restoration of critical operations within _____ hours.
3. To provide for the restoration of all operations within _____ days.
4. To achieve the foregoing in a cost effective manner.

APPENDIX 5

EXAMPLES OF POSSIBLE SUPPORT TEAMS

Survival Control

This team can perform an intermediary function between support teams and functional teams. It includes the responsibility for coordination of the company's internal information flow, maintaining a current recovery status of all functions displaced by the crisis. This information will be communicated to other areas which would normally interact with the displaced functions. This team maintains the central command centre which will be the communication link between the division management team(s), support teams and functional teams.

Evaluation

The team assigned this function will assemble at the disaster site to evaluate the situation. Information will be gathered and reported to the division management team(s) in an initial evaluation report. The information will be such that the division management teams will be able to make the short term strategic decision of whether to stay in the damaged facilities and operate on a degraded basis or to move operations to an alternate site. This report should be completed within 3 to 4 hours of the disaster. Following the completion of the initial evaluation, an in-depth assessment of the damage to the facilities will be made and the information detailed in a final evaluation report to the division management teams. This report should include an estimate of the length of time that the facilities will be out of service so that the division management teams can make long term strategic decisions. It should be completed within 3 to 4 days of the disaster.

Salvage

This function is for the coordination of salvage operations and the repair of items contained in the damaged area. An inventory of useable equipment, furnishings, documents and supplies will be provided to the team assigned the procurement function. They will determine the salvageable supplies and materials and arrange for their removal, relocation or repair. It will (with the appropriate vendors) prepare an equipment inventory estimating the condition of each item. Communication with the survival control team is to be maintained so that changes in equipment status and schedules are updated.

Facilities

This function is for the assembly, organization, and maintenance of the alternate site requirements for all functional areas. It also has the responsibility for maintaining contacts with local estate agents to identify available facilities which could be used as alternate sites and for working with the team assigned the procurement function, to obtain use of the facilities when the need arises. Coordination should be maintained with the team(s) assigned the communications functions to ensure that sufficient communications capabilities are available in the proposed facilities.

Physical Security

This function is for the provision of security at the temporary sites as well as the permanent site. This will include providing security for personnel and securing physical facilities from unauthorized access. They may also be used to help transport resources and personnel between sites.

Voice Communications

This function is for coordinating the restoration and/or establishment of minimum requirements for voice communication lines and equipment with appropriate vendors. Coordination should also be maintained with the team assigned the facilities function to ensure that sufficient voice communication capabilities are available in the possible alternate facilities. Coordination with the team assigned the procurement function may also be required.

Data Communications

This function is the coordination of restoration and/or establishment of minimum requirements for data communication lines and equipment with appropriate vendors. Coordination should also be maintained with the team assigned the facilities functions to ensure that sufficient data communication capabilities are available in the possible alternate facilities. Coordination with the team assigned the procurement function may also be required.

Procurement

Under this function, the immediate provision of the minimum quantity of furnishings, office and production equipment, supplies and forms to support those functions affected by the disaster is accomplished. Coordination with the teams assigned the functions of facilities and shipping/receiving/distribution will be required.

Shipping/receiving/distribution

This function is for coordinating the internal movement of resources required by the survival effort from vendors, off-site storage and 'spares' sources to and from the alternate sites.

Travel

This function will address the provision of transportation and accommodation for personnel. This will entail gathering the transportation need from the survival teams and then identifying and arranging with the various transportation providers the acquisition of emergency transportation. Primary and alternate means will be identified. The existing policies on travel restrictions and travel insurance should be taken into consideration.

Fleet Administration

This function is to provide vehicles normally used by the organization in their fleet pool (cars, vans, trucks, etc.) for the survival effort.

Records Management

This function entails the coordination of retention and retrieval of records necessary for survival. It also includes coordination of storage, retrieval and delivery of survival resources which must be stored off-site.

Finance

This function must establish policy and procedures, for

the emergency approval and accounting of survival expenditures, before a disaster. It will also have the responsibility of overseeing the emergency funding procedures after a disaster.

Personnel

This function will oversee the provision of human resources needs. This may include the provision of qualified manpower to fill vacancies in the functional areas caused by absence, injury or death, from existing organization rosters or temporary personnel sources.

Public Relations

This function will establish policy and procedures for internal and external communications during an emergency, with the approval of executive management. It will then be the single official information source during the recovery.

Mail Services

This function is the re-establishment of both internal and external mail servcies to affected functions at their alternate sites. It also coordinates the restoration of courier services to affected areas.

Medical Services

This function will ensure that 'First Aid' services are provided to the victims of the emergency.

Insurance Services

Coordination of insurance services is the responsibility of this function. Special emphasis will be made on employee health and death benefit insurance and property and fire insurance. It will coordinate with insurers to assure compliance with their requirements in order to achieve maximum return. Coordination with the team assigned the finance function will be required to ensure that expenditure accounting procedures are adequate for insurance purposes. It will then oversee expenditure accounting during the recovery to assure proper insurance reimbursement to the organization. Continuous coordination with the support teams assigned the facilities and procurement functions will be necessary to ensure appropriate limits of liability are maintained for overall extra expenses coverage.

Legal Services

This function will maintain familiarity with pertinent clauses in contracts concluded by the organization in order to provide advice during the survival effort on such things as non-performance clauses, etc.

Micrographic Services

This function will coordinate the production and storage of critical micrographic documents in an off-site location. It will arrange for the re-establishment of minimum micrographic services following a disaster, as well as coordinate the delivery of critical documents to the affected functions at their alternate site.

Microcomputer Support

This function will address the acquisition of equipment, software and documentation associated with lost, essential, microcomputer resources.

Word Processing Services

This function provides overall word processing services to support the survival process, as well as secretarial support as required by the other support teams.

Reprographic Services

This function replaces the inventory of forms the organization produces in-house and which are critical to the affected functions.

APPENDIX 6

SAMPLE BUSINESS FUNCTION ANALYSIS FORM

Completion Procedures

I. Introduction

This procedure is provided to assist in the understanding needed to prepare the functional analysis form. The information collected on this form will be used in the setting of priorities and impact analysis, detailed interviews and in the writing of the survival plan.

II. Filling Out The Form

   A. Headings

   1. Function Performed

   * Survey all functions in your area of responsibility. Each function should have a form filled out for it.

   Example:*Accounts Payable*

   2. Description of Function

   * Give a brief description of the function.

   Example:*Pays invoices, expense accounts and accounts for monies owed by the organization*

   3. Department Name

   * Give department name.

   Example:*General Accounting*

   4. Department Head

   * Give name of department head

   5. Reports to

   * Give name of person and organization the department reports to.

   Example:*Finance, Mr. XXXX*

   6. Is this function automated?

   * We want to know if it uses automated support and if so to whom does it belong? Could it be done any other way if this support were not available?

   7. Are there critical dates or periods associated with this function?

   * Tell us what dates, if any, are important.

8. Is this function required by law, contractual obligation, or regulatory reporting?
   * What must be done because of legal or other regulatory requirements?
9. Priority/Frequency
A. Prioritize this function by placing a circle around the correct category below:
   * By itself
     Ask three questions as to the priority it must have if a disaster would occur, do not try to take each function and compare it to other functions. Just compare it to the three questions and circle the numbers of the one that fits, remembering that we are asking for organizational priority.
B. Circle the appropriate category to indicate the frequency the organization should establish for this function during recovery.
   * Look at each function by itself. How often must we perform this function if a disaster should occur?
10. Estimated revenue earned by the organization from this function is (currency) per quarter.
    * From a monetary stand point, how important is this function?
11. Estimated cost or loss to the corporation if this function were not performed.
    * What impact would time have if this function were not performed?
12. Other intangible losses if not performed.
    * Losses, other than monetary, that would be incurred if this function were not performed.
13. Special considerations or comments.
    * Tell us why this should/should not get done — we want to know more about it.
14. Prepared by:
    * Individual who filled out the form and telephone number.

    _____

    * Phone number where you can be reached.

    _____

    * Date form completed _____

  B. Make a copy of all forms for your records.

Business Function Analysis

I. Function performed _____
_____

II. Description of function: _____
_____

III. Depament name:_____

IV. Department head:_____

V. Reports to:_____

VI. Is this function automated? YES ____ NO ____

  A. If **NO**, proceed to VII.

  B. If **YES**, does it use services supplied by Central Data Processing?
   YES ____ NO ____
   If **YES**, proceed to 'C', if **NO** please explain and proceed to 'C'.

   _____

   _____

  C. Is manual processing of this function possible if Data Processing service is lost?
   YES ____ NO ____ If **YES**, please explain.

   _____

VII. Are there critical dates or periods associated with this function?

   Date/Period   Explanation

  1. _____

  2. _____

  3. _____

  4. _____

VIII. Is this function required by law, contractual obligation, or regulatory reporting?
  YES ____ NO ____ if **YES**, please explain.

   _____

IX. Priority/Frequency

  A. Prioritize this function by placing a circle around the number of the correct category below:

1. This function is absolutely essential for the organization to remain operational.
2. This function is not critical, but should be performed as early as possible after the department is fully operational or if there is time available upon completion of category one functions.
3. This function is not necessary for the organization to remain operational.

B. Circle the appropriate category to indicate the frequency the organization should establish for this function during recovery.

Daily   Weekly   Monthly   Quarterly   Anually   As Required   Other

X.   Estimated revenue earned by the organization from this function is (currency) _____ per quarter.

XI.   Estimated cost or loss to the corporation if this function would not be performed.

One day _____   Two days _____
One week _____   Two week_____
One month _____   Longer_____
Currency used: _____

XIII. Special consideration or comments:

_____

XIV. Prepared by: _____   Phone: _____
Department: _____   Date: _____

APPENDIX 7

SAMPLE DATA PROCESSING APPLICATION ANALYSIS FORM

1. Application Name _____
2. Functional Description of Application*

    _____
    _____
    _____

3. Batch ____ Online ____ Inquiry ____ Update _____
4. Primary User, Department, and Contact

    _____

5. If input is online, can batch input be used?
    YES ____ NO ____
6. Can the input from two or more cycles be combined, allowing the application to be processed less frequently?
    YES ____ NO ____
7. If telecommunications is used for I/O, can other means (e.g., mail, courier services) be used?
    YES ____ NO ____
8. If telecommunications is used for I/O can user activity be scheduled during specified time slots so that available resources may be shared among applications?
    YES__ NO__

9.A. Prioritize this application by placing a circle around the number of the correct category below:
    1. This application is absolutely essential for the organization to remain operational. A loss of this application would have a **major** impact on organizational operations.
    2. This application is not critical, but would have some impact on organizational operations.
    3. This application is not necessary for the organization to remain operational, and has **little** or **no** impact.

B. Circle the appropriate category to indicate the processing frequency that the organization should establish for this application during the recovery.
    Daily   Weekly   Monthly   Quarterly
    Anually
    As Required

10. Special considerations or comments: _____

    _____

11. Prepared by_____   Date_____

SAMPLE DATA PROCESSING APPLICATION ANALYSIS FORM

Department_____ Telephone _____

* Please describe in terms which will be understandable by the organizational users of this application. Include major reports generated by this application.Data Collection list for the Survival Coordinator

1. COMMAND CENTRE LOCATIONS (Form provided)
   List corporate facilities and hotels/motels where temporary headquarters could be established following a disaster. Locations at radii of 2, 5, 8 and 15 kilometres from the primary facilities should be included.

2. INSURANCE COVERAGE INFORMATION (Overview provided)
   Collect copies of insurance policies which provide coverage for the areas listed below. Copies of the full policy with endorsements are required.
   a. Data processing equipment
   b. Magnetic media (e.g., tapes, disks)
   c. Extra expense
   d. Business interruption
   e. Computer crime
   f. Data processors errors and omissions.

3. NOTIFICATION INFORMATION FOR MANAGER RESPONSIBLE AS THE DESIGNATED SUPPORT COORDINATOR. THE FOLLOWING INFORMATION IS REQUIRED FOR EACH;
   Name:
   Title:
   Address:
   Postcode/City:
   Home Telephone:
   Office Telephone:

APPENDIX 8

SUGGESTED CONTENTS OF A DATA PROCESSING SURVIVAL PLAN

A — Introduction
* How to use the Survival Plan
* Survival Plan Synopsis
* Numbering Scheme Explanation
* Policy Statement
* Project Foundation
* Project Objectives
* Method of Distribution

B — Duties of the Survival Coordinator
* Overview and Summary
* Survival Training Programme
  * Summary
  * Orientation Training
  * Team Training
  * Training Schedule
  * Testing Programme
    * Summary
    * Component Testing
    * Integrated Testing
    * Disaster Drill Testing
    * Testing Schedule
  * Maintenance Procedures
    * Summary
    * Ongoing Maintenance
    * Maintenance Responsibilities
    * Maintenance Schedule
  * Functions During Recovery
    * Summary
    * Plan Activation
    * Establishing a Command Centre
    * Advisor of Survival Teams

C — Data Processing Survival Teams
* Synopsis of Survival Teams
* Teams:
  * Recovery Strategies
  * Recovery Procedure/Progress Log
  * Primary and Alternate Membership
  * Task Assignments
  * Task Procedures

* Support Resources including:
    Organization Chart
    Glossary of Survival Terms
    Synopsis of Support Functions
    Priority Definitions
    Production Application Priorities
    Command Centre Locations
    Calling Procedure and Telephone Log
    Transportation Request Form
    Recovery Status Report
    Production   Application   Processing
    Constraints
    Production Application Support
    Responsibilities
    Recovery File Requirements
    Evaluation Checklists
    Critical Services
    System Component Failure Impact Analysis
* Contact Lists for:
    Alternate Sites
    Users/Customers
    Staff
* Vendors and Servicers of:
    Hardware
    Telecommunications
    Software
    Office Equipment
    Forms & Supplies
    Furnishings
    Air Conditioning
    Utilities (Power & Water)
    Manuals
    Off-site Storage Facilities
    Alternate Services
* Inventory Lists for:
    Forms & Supplies
    Manuals
    Software
    Office Equipment/PCs

CONTINUITY PLANNING

APPENDIX 9

SAMPLE TRAINING REQUEST FORM

Submitted by: _____ Date: _____

Job title: _____ Dept: _____

Tele ext._____ Training request for

Individual _____ Group _____

If group, number of people: ____

Individual/group training requested for: (circle one)

    Orientation

    New team member

    New/updated information

    Test preparation

    Alternate team leaders and members

Requested training date (give three alternative dates)

    1. _____

    2. _____

    3. _____

Person to contact to verify date: _____

Department: _____ Tel ext: _____

Additional information or requests: _____

_____

_____

_____

_____

_____

RETURN TO SURVIVAL COORDINATOR

APPENDIX 10

SAMPLE POST TEST QUESTIONNAIRE

In order to evaluate the effectiveness of the survival plan test, please take a few moments to answer the following questions. Your comments will help improve future testing activities. Please leave blank the questions which do not apply.

YES   NO

1. Were test objectives clear?
    Comments: _____
    _____

2. Were scenarios, if used, adequate in providing direction for test activities?
    Comments: _____
    _____

3. Were other test materials helpful during test activities?
    Comments: _____
    _____

4. Did the test help you understand your function during recovery activities?
    Comments: _____
    _____

5. Did the test help you better understand the survival plan?
    Comments: _____
    _____

6. Did this test give you an opportunity to adequately interact with other team members and teams?
    Comments: _____
    _____

7. Are there parts of the plan, used during the test, which you feel are not clear?
    Comments: _____
    _____

8. Was there information you required during the test which was not in the plan?

Comments: _____

_____

9. Please rate the effectiveness of this test exercise in giving you a better understanding of your recovery duties (circle one):
GOOD   FAIR   POOR
1   2   3   4   5
Comments: _____

_____

10. Additional Comments: _____

_____

Name (optional): _____ Extension _____

APPENDIX 11

EXAMPLE AUDIT CHECKLISTS

Section III
A: Introduction
YES   NO

1. Is there   formal introduction to the survival plan?
Comments: _____

2. If so, does the introduction include:
a. Details concerning the use of the plan?
b. Details   concerning   the organization   (numbering scheme) of the plan?
c. A synopsis of the plan?
d. Method of distribution and control?
Comments: _____

3. Are the objectives of the survival plan clearly stated?
Comments: _____

4. Is there a policy statement detailing   management's commitment to the plan?
Comments: _____

5. Has an individual been assigned to   coordinate   control, maintenance, and testing of the plan?
Comments: _____

6. If an individual has been assigned,   have   his responsibilities been itemized and included the plan?
Comments: _____

Section III
B: Corporate
Considerations
YES   NO

1. Are any required support areas included as part of the survival effort? For example:
    a. Finance
    b. Payroll
    c. Employee Insurance
    d. Building Insurance
    e. Legal
    f. Purchasing
    g. Security
    h. Mail Services
    i. Personnel
    j. Transportation
    k. Public Relations
    l. Evaluation
    m. Communications
    n. Facilities
    o. Records Management
    p. Disaster Control
    Comments: _____
    _____

2. Have their responsibilities been defined?
    Comments: _____
    _____

3. Are these areas familiar with their responsibilities in the survival effort?
    Comments: _____
    _____

4. Have responsibilities, tasks and resources for executive management been identified and documented in the plan?
    Comments: _____
    _____

B.1

143

APPENDIX 12

EXTRACTS FROM AN ACTUAL PLAN AUDIT REPORT

The following is an extract, quotation, from an actual report following a plan audit for a major customer — suitably editted to disguise the client.

The extract includes Section 1, the Conclusions of the report, Section 2 'the Recommendations and a few pages from Section 3' the Summary Matrix.

1. CONCLUSIONS

The client has invested a significant amount of money and time in the task of reconstructing its present operational data processing systems at the alternate site. The final goal of obtaining a continuity plan for the Automation Division has, however, not been reached. The main reason is the unstructured approach that has been taken. Responsibilities have not been clearly asssigned and objectives have not been issued, consequently there is no overall coordination and management of the survival programme.

While in our belief continuity planning is a top down activity, the client separated the activities in this area from the normal day to day operations and organization structure. One result of this is that the staff do understand the necessity for a complete and tested continuity plan and are demotivated by the apparent lack of success. The existing plan is incomplete and not accurate in many areas. The details can be found in section 3, but below we provide a summary of our major observations.

A. Management, Organization Issues

- The plan does not indicate what the guaranteed recovery times are in the following areas:
  * Automation Division towards the end user.
  * Supplier towards the client on network recovery.
  * Alternate site towards the client on machine availability in a real disaster.

- The manual only involves users who are (Telephone Sales) or were (Logistics) located in the computer centre when the survival project started some five years ago. Plans for end user computing tools are omitted.

— No attempt to evaluate what response times are acceptable in an emergency situation has been carried out.A formal way to evaluate test results with the alternate site does not exist, this leads to misunderstanding and 'finger-pointing' at lower staff levels.

B. Technical Issues

— Major problems are encountered in testing as no formal procedures are in place to keep the production systems in synchronisation with the alternate site systems. Where procedures exist, they are not closely followed.

— Tests at the alternate site indicate that some systems at this moment could be recovered in a technical test environment. No user activity has ever been tested or simulated. The area of transaction load on to the network has never been tested.

— No plans or priorities have been made for printing the high volumes of daily printed output. Printing capacity at the alternative site is limited and incompatible with the laser printers used by the client.

C. Economics

— The client has obviously committed significant financial and manpower resources to the alternate site activities over a period of five years. There is however, no mechanism in place to fully evaluate the total cost or the return on this investment. The lack of this type of information has led several members of the client's staff to question the value of this whole exercise.

— The number of man years spent annually is at least six but could be as high as ten. This is spent across several departments within the Automation Division.

D. Status of Present Plan

— The client's plan is not a continuity plan in the classic definition. It actually describes a test environment in which some groups of people are executing a 'technical' exercise of recreating an operational computer environment on a different piece of hardware.

- Part of the plan is not completed and is still in project status. Several man years of effort have to be spent before it is complete.

- The data in the plan is incomplete and often out of date.

 * Staff members no longer employed by the client are still mentioned.
 * Employees have other responsibilities than those mentioned.
 * Most staff members we interviewed knew about the plan but did not have a copy, had not read it or were not even aware that their names were in the plan.

* Support areas outside the Automation Division but inside the client organization were not sure if they could provide the requested service, such as transporation, in a real crisis situation.

* Support areas outside of the client such as the vital PTT were not aware of any tasks they would have to carry out. No formal arrangements are established with the exception of those for the alternate site.

* Only one team is clearly identified in the existing plan. This is the 'Crisis Team' which consists of the four most senior managers of the Automation Division.

 * One of these members was not aware he had a responsibility in this area.
 * How the plan would be activated if none of these managers are available, is not clear.

E. Security

Security both in the area of physical and information needs improvement.

* Access to building. Doors are left open and procedures for entry are not enforced.

* Off-site tape/cassette storage. The building is in a non secure area. Entrance to the storage room is not protected other than by normal type office door and lock.

*   The electrical power supply is an easy target for sabotage or terrorist action, because it is located near the fence and clearly identifiable.

These are just observations. We did not carry out a full security review, therefore this summary is only an indication that security should be improved.

2. RECOMMENDATIONS

The recommendations that follow are grouped into short term and long term categories.The short term recommendations address the problem of the client's inadequate survival plan and propose a means of improving that plan to make it workable.

The long term recommendations review some important points which should be included in the client's strategic planning for disaster recovery.

2.1. Short Term

The client should develop a more comprehensive plan using a top down approach, formalizing policy and the objectives of the plan, prioritizing applications and forming teams to address all of the functions required to effect a successful recovery.

We summarize below specifically how the plan should address the inadequacies under the same topic headings used in Section 1, Conclusions.

A. Management Issues Organization

The plan should become an integrated part of the organization's activities so that, when tasks and responsibilities are assigned to individual managers, they can control the continuity plan activities, identify problem areas and take corrective action, if needed, in a timely manner. During the development of the plan, if the iterative process is followed, the staff are trained to execute their tasks in the survival activity with confidence and this increases staff motivation. Management, being involved and responsible, will have the tools to execute their responsibilities in the recovery procedure. Users will eventually be involved so that they understand and accept the objectives and characteristics of the recovery for which the plan is developed.

B. Technical Issues

The project should include the development of testing

procedures which will take advantage of the alternate site's expertise and input. These procedures will provide identifiable responsibilities to appropriate management to ensure that methodical and organized testing is accomplished. This will allow the testing process to have the best opportunity for success.

C. Economics

A more organized method of testing should allow a reduction in the number of tests each year. It is probable that the present frequency could be reduced to approximately four, thereby reducing those associated costs. We do, however, recommend tests which fully reflect the hardware situation as it will be in a real recovery situation.

D. Status of Present Plan

The present plan is incomplete and out of date. The new plan should ensure that each operational unit is examined. From this the survival and support teams as well as the inter-team communication procedures should be developed. Team tasks should be documented, supported by required resources listings and detailed procedures. The teams that help develop the plan should be the same ones that will also direct the recovery. Special sections in the plan should be dedicated to training methodologies and plan maintenance. The maintenance methodologies should call for defined and comprehensive maintenance tasks to be specifically assigned to individuals.

E. Security

A full Security Review and Audit, should be performed as an integral part of the continuity plan, to help identify security exposures on a priority basis.

We recommend that a fully documented plan be developed using the application priorities and recovery time-frames as they are defined at present. In the planning process, confidence in the ability to actually recover within the estimated time-frames will be established. Then it is recommended that an impact analysis be conducted to confirm both the accuracy of the priorities and the adequacy of the recovery time-frames with the users of the Automation Division's services. Any changes that appear necessary can then be more easily integrated by adjusting procedures and agreements in an already functional plan. Although this sequence is somewhat different from that normally used in a continuity planning project, it would be more effective for the client circumstances.

2.2. Long Term

Continuity planning must be part of day to day operations for the plan to remain a 'living' document, a reflection of the living and changing organization. Continuity planning must be a part of strategic planning so that long term development can include effective recovery capability. Some important points to be considered are as follows:

— Should the client continue with the present alternate hotsite methodology?
— What alternatives are available?
  — other hotsite suppliers
  — build an in-house solution by having two data centres
  — hardware and software consequences
  — data communication development
  — other outside solutions (other similar organizations)
— What are the financial consequences?
— What is expected of the Automation Division by the users?

3. SUMMARY MATRIX

Attached* you will find our summary matrix in which we compare the client's existing plan with the recommended standards.

In previous sections of this report we have already commented on the fact that in our view the client does not have a survival plan which ensures that all tasks are documented, teams are created and trained ahead of time. The matrix covers 93 items which could be further exploded into several hundreds of details. However, if the answer to the top level question of a section is negative, or negative to a large extent we do not explore the next level of detail for that section. Details are in section 4 of this report.

Recommended standards

Present	4	4.3%
Present but not complete	12	12.9%
Not Present	77	82.8%
Total standards	93	

The clearly illustrates that the current client plan is very limited in scope and scale, and is really a technical status rather than an executable action plan.

* Authors note:I have included some pages of this matrix as an illustration.

Summary Matrix
COMPARISON WITH RECOMMENDED STANDARDS

PREFACE/INTRODUCTION Subject	Present	Not Present	Not Complete	Comments
How to use the EDP Operations Disaster Recovery Plan			X	
Disaster Recovery Plan Synopsis			X	
Numbering Scheme Explanation			X	
A division policy statement, signed Policy Statement			X	A division policy statement, signed off by executive management can significantly enhance ongoing support of the Survival Plan.
Project Foundation			X	
Project Objectives			X	
Method of Distribution		X		
DUTIES OF THE DISASTER RECOVERY COORDINATOR Subject:				
Introduction			X	
Disaster Recovery Training Programme			X	
Overview			X	We encourage development
Orientation Training			X	of a training program which
Team Training			X	will ensure that awarenes
Training Schedule			X	of the plan is corporate wide through Orientation Training and changing team structure or membership.
Disaster Recovery Testing Program				
Overview			X	While testing of the plan
Component Testing			X	has been ongoing, we feel
Integrated Testing			X	that a thoroughly documented
Testing Matrix Disaster Drill Testing			X	testing program will ensure increased success and heightened awareness of the recovery process.
Testing Schedule			X	
Maintenance Procedures			X	
Overview			X	Maintenance of some portions
Ongoing Maintenance		X		of the plan has occurred.
Maintenance Matrix			X	A more formalized
Maintenance Schedule			X	maintenance programme will ensure that the plan is an accurate reflection of EDP operations.
Functions During Recovery Effort				
Overview			X	
Plan Activation		X		
Establishing a Command Centre	X			
Advisor of Recovery Teams			X	
DATA PROCESSING DISASTER RECOVERY TEAMS Subject:				
Synopsis of Recovery Teams			X	The reliance of the plan
Teams				on the availability of an

CONTINUITY PLANNING

Flowchart	X	existing management is a.
Overview	X	weak point. We believe
Primary and Alternate Membership	X	that the development

of teams, based on corporate
structure before a disaster
occurs will ensure a
smoother recovery operation.
Many of the points reviewed
in this portion of the checklist
could easily be expanded.